THE

NEW

FARMER'S
ALMANAC

FOR THE YEAR

2013

GREENHORNS

A Note

by SEVERINE VON TSCHARNER FLEMING
EDITOR-IN-CHIEF

We have compiled this Almanac under one guiding premise: that the bravery required of new farmers today requires history as a context. We hope that we may continue our efforts in this direction, and we welcome the involvement of others in the continuation of the Almanac idea.

Our work with Greenhorns over the past five years has been to convene young farmers, to support them with education and community, to promote the ideas of this movement in the mainstream press, to document their struggles and advocate for solutions, to be a hub of solidarity and a network for peer-to-peer learning. How strange to find ourselves as historians and micro-publishers! We started as new media kids: indie documentary, blogging, internet radio, pop-up mixers and press stunts, and now we're sticking our toe into old media, as green as ever. I hope this organization can stay nimble and flexible in the coming years, and take on many expressive forms. We endure because of your support and we welcome your involvement.

In my introduction I set out some key markers in the early part of American agricultural thinking--as useful landmarks and to guide your study, I am only an amateur historian, but this yearly format for tackling new material seems workable as a learning tool. Reflexively then, this read of history considers my own practice of sustainable agriculture and community organizing-- and how it relates to the great issues of our time; how it has always related to the great issues of all other times. Big dreamy minds have latched on to these themes of soil, of startup, open source, direct action, local politics, economic transformation, and indeed these may well be the shapes and themes most suited to shifting the systems. we inhabit. The challenge therefore is to use history as a tool, to sharpen our wits, to steel our resolve-- and to forsee big changes of our own making.

CALCULATED FOR THE YEAR 2013 BUT GOOD FOR ANY YEAR.

CONTAINING A GREAT VARIETY OF USEFUL AND ENTERTAINING PIECES PERTAINING TO YOUNG FARMER IN AMERICA.

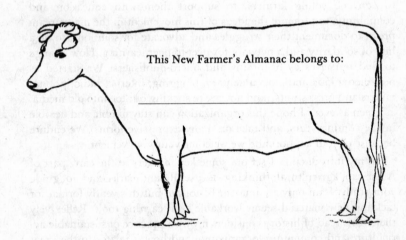

This New Farmer's Almanac belongs to:

_____

Introduction
by SEVERINE VON TSCHARNER FLEMING

Affixing our activities on the zodiac seems anachronistic nowadays-- the stars are so ancient, we think. Those celestial beasts -- Canus, Taurus, Lupus, Ursus -- may have cast their spells on the weather for eons, but can those predictions persist in a sky contaminated by global capitalism? Is the old advice still worth seeking? Can the agrarian idioms we've gleaned by living a life that touches nature, from old books, and from the vicarious viewpoints of our neighbors-- can those values hack it alongside suburban sprawl, jet-set digitalism and a distorted food economy? Are our aspirations reasonable? How should a contemporary agrarian conceptualize the shift ahead, and our role in it.

Admiring the hand-hewn Chestnut beams and floorboards in the barn we are restoring, their wise placement on the hill, I noticed how the breezes flood in, and I wonder: "Would I have had the good sense to build this? Probably not." We grope with big themes as we pursue this ancient practice of agriculture, and though the mainstream society practically ignores this knowledge, it seems we cannot learn enough to unlock yet more potential from our farm and land and place. Under the sky, under the stars, we have time (what a priviledge) to consider our lives and what we can do with them. It is possible to quiet the mind, especially with so much time screwing and unscrewing hoses, moving fence, watering seedlings with regular swoops of the sprinkling wand. In this way, stillness and reflection coexist with routines and chores, observation brings in new themes of inquiry, every day is a catalog of small, useful insights, which we can attach to visual cues alongside the inventories of grain, hose, bits and valve, and brainpower to design our own personal theories of change.

The farm provides associative context for many kinds of knowledge, metaphorical and botanical and social, mechanical and institutional. Sometime it's as plain as the joinery in the hay mow, the corral and the shute. Others are complex: the fruiting tree with its pollinators and frost dates, the Grange with its moral purpose and potluck glue, the sugarbush with sap rising, snow underfoot for the sleigh, the Farmers Union and its populist education arm, the CSA, which give us a scaffold to build commerce, community, and a more tempered cash flow in our seasons, the neighbor's sileage crop, with its dose of GMO shrapnel along the roadsides, to rodents and muddy runoff. The landscape in which we are working is a system. Our inputs and actions affect whole systems, not parts of them. Our actions have practical implications on a thousand such overlapping systems, natural, agricultural, social and economic, striving always to make manipulations that will support our goals, our soils, and our bottom line. We are dealing with seasons as they come , and if we're good, every decision around the farm can bring confidence that our actions will add up.

That is the purpose of this essay -- to see if history can give us some clues about what might be possible, before we are so quick to dismiss the spirit, the stars, the sacred land practices, the agrarian traditions, the oldy-timey knowledge, the community values that are bound up in a kind of complicated and contradictory bundle of topics that adjoin agriculture. How do these practices on the land translate into the entire human system? How much can we hope for? What have others achieved before us? I wonder about these things just as the builders of our barns must have done, just as the builders of our nation must have done.

Part I

A HISTORY OF LOOKING BACK

As this amateur historian has discovered, keeping time from observation of the sky and the sun as a way of ordering human actions is far older than agriculture itself. Knowledge of natural sequences was in fact a prerequisite to farming. " Native people drew spiral pictographs - sets of connected concentric rings radiating out from a center -on cave walls and rock shelters in locations where they are illuminated by the rising sun on the winter solstice...In this tribe were some older people who set up signs to gauge how the sun shone. They found they had to keep moving the stick to the right for a long time, and then to the left, and so by this means they discovered what times the birds had their nests, and what times the animals had their young, also what times the plants grew , and the time the seeds were ripe. This they did year after year as they studied the signs of the sun." The first calendars were created by indigenous clockmakers, (of course!) who kept time to their observations, integrated instinct, conventions, experience and ritual, all in an eco-logically durable and highly spiritual culture bent on "tending, but not taming the wild."

"The physiological and behavioral changes that occurred in animals during seasonal and even daily cycles were also crucial to Indian cultures. Animals were harvested in certain seasons, or when migration cycles caused natural congrega-tions. Rabbits were taken when fattest, fish when spawning, birds when molting, crickets in the cool mornings when they bunched in the grass" We may think of agriculture as the 'break' from indigenous land practices, but this is not true. What is true is that farming for indigenous prac-titioners is not secular, but rather indivisible from spiritual practices and the biocultural context of sacred plants and sacred places. The land ethic that guides it is highly principled and is ordered by tradition. There are still some places left in the world where indigenous peoples manage their landscapes in this manner.

Next, I hit the books to try to find a golden era in agriculture, an era of peaceful interface between humanity and ecology-- maybe not quite as pure as the sacred, indigenous way, but perhaps a more approachable intermediate step. There must be some 'golden examples' of farming that works well for those involved as well as the landscape, right? What I found were the writings of other hopeful seekers also looking back. It's a theme perhaps as old as agriculture itself! In Virgil's famous epic poem " The Georgics," he describes the culture and care of olives, wine grapes, and the breeding of animals in the Italian hill country of 29 BCE. It was a permanent, perennial, adapted and sound agriculture that fed the early Classical world, the fantasy landscape for today's Permaculturists and Tree Crop enthusiasts. But Virgil, too, mourned a golden time, just past, when...

> Fields knew no taming hand of husbandmen
> To mark the plain or mete with boundary-line.
> Even this was impious; for the common stock
> They gathered, and the earth of her own will
> All things more freely, no man bidding, bore.

Virgil's poetic descriptions do bear some traces of inferred relationships between plants and people, though he is explaining the science of grafting vines. But it's interspersed with 'natural law' and rules that work. Gods must be satisfied, beasts must not be abused, plants must be treated just so--" Lest land lie Idle". Again there is an order to things, and wildness, spirit and culture intermingle freely.

FOOTNOTES

Pg. 51, 61 Tending the wild, Kat anderson, University of California Press, 2005
Georgics, Book 1: 125-28

Virgil describes a 'lost time,' a pastoral landscape less dominated by the forces of the market, the hunger of the empire's armies, where the quiet yeoman farmer gathered and ranged with his happy beasts, un-enclosed, peaceful and self-sufficient in harmony with his neighbors and with nature. It seems a long standing agrarian ache, to harken back to a purer time, a time with less exploitation, less confinement, less extraction and servitude.

Why do we have this constant pattern of retreating to rustic arts, of " back to the land" utopianism ? It is a frequent and persistent theme in American history. Being an American farmer, I'll start there. The earliest text I could find was Nathaniel Hawthorne's retreat to Brook Farm in 1852, fictionalized in his book, The Blithedale Romance, which traces a failed but poignant effort at a more natural way of living, a ' thought experiment' by a salon of highly privileged intellectuals tossing around big ideas while milking cows. Frankly, most of the early settlements were guided by moralistic and spiritual quests for purity and justice in this country, but even for Hawthorn Nature was more often termed as " wilderness" and devilish than as a source of natural order or spiritual value.

Some might try to dismiss these "Blithedale Romanticists" and their kind as holding a minority viewpoint, the " loser" of history -- for indeed we live in nation whose agriculture is almost totally industrialized, rationalized and corporate. But this thread of agrarianism has endured it all, with the romantic "Old Macdonald had a Farm" holding strong against all odds. Commercial/ commodity farming may dominate the landscape, but in our imaginations and mountain valleys a different dream persists, a dream wrapped up in a broad set of moral values, democratic principles. Indeed the broad scope of human enterprise is well represented in this literary tradition. If you are a new student of agrarian history, like myself, I urge you to start start at the beginning and work your way through. I spent far too long with my nose in the Gardening section. Since finding the History section I am a changed person.

To convince you, Dear Reader, to delve into this study for yourself, I've attempted here to lay out some good themes and titles so that you can start your own course of study from the cheapest books on Abe.com or the local used bookstore. Or even read these titles online at archive.org.

Here goes: Late 18th and early 19th century America (aka the industrial revolution) saw the fermenting genesis of dozens of reform-minded experimental farms, many of them responding directly to the disturbing configurations of power in the mill towns, the smoke, the hours, the corruption. Upstate NY , where I live, was a spawning ground for such upstarts as productive landscape at once close enough to the affected populations and market-power yet still ripe with opportunity for those who would claim it, a frontier for wildness-craving American philosophers.   Bostonians Emerson (Nature, 1836) and Thoreau

First, nature's law
For generating trees is manifold;
For some of their own force spontaneous spring,
No hand of man compelling, and possess
The plains and river-windings far and wide,
As pliant osier and the bending broom,
Poplar, and willows in wan companies
With green leaf glimmering gray; and some there be
From chance-dropped seed that rear them, as the tall
Chestnuts, and, mightiest of the branching wood,
Jove's Aesculus, and oaks, oracular
Deemed by the Greeks of old. With some sprouts forth
A forest of dense suckers from the root,
As elms and cherries; so, too, a pigmy plant,
Beneath its mother's mighty shade upshoots
The bay-tree of Parnassus. Such the modes
Nature imparted first; hence all the race
Of forest-trees and shrubs and sacred groves
Springs into verdure.

Other means there are,
Which use by method for itself acquired.
One, sliving suckers from the tender frame
Of the tree-mother, plants them in the trench;
One buries the bare stumps within his field,
Truncheons cleft four-wise, or sharp-pointed stakes;
Some forest-trees the layer's bent arch await,
And slips yet quick within the parent-soil;

No root need others,
nor doth the pruner's
hand

   Shrink to restore the
topmost shoot to earth

   That gave it being.
Nay, marvellous to tell,

   Lopped of its limbs,
the olive, a mere stock,

   Still thrusts its root
out from the sapless
wood,

   And oft the branches
of one kind we see

   Change to another's
with no loss to rue,

   Pear-tree transformed
the ingrafted apple yield,

   And stony cornels on
the plum-tree blush.

   Come then, and learn
what tilth to each belongs

   According to their
kinds, ye husbandmen,

   And tame with culture
the wild fruits, lest earth

   Lie idle. O blithe to
make all Ismarus

   One forest of the
wine-god, and to clothe

   With olives huge
Tabernus! And be thou

   At hand, and with me
ply the voyage of toil

   I am bound on, O my
glory, O thou that art

Georgics Book 2

   Upon the dying beast;
the skin is dry,

   And rigidly repels the
handler's touch.

   These earlier signs
they give that presage
doom.

   But, if the advancing
plague 'gin fiercer grow,

   Then are their eyes
all fire, deep-drawn their
breath,

   At times groan-
laboured: with long
sobbing heave

   Their lowest flanks;
from either nostril
streams

   Black blood; a

(Walden, 1854) got the ball rolling, but there were many less literary types out hoeing. The Mormon startup (1817) in western NY was one of many fledgling religious movements, and it succeeded. The Mormons marched off across the plains in search of their freedom, then across deserts, finally settling in Utah and establishing farms, mills, plantations of mulberry trees for silk production, and vast farming syndicates that survive to this day. The Shakers, also in NY, on what is now the Albany Airport, led by a founder Ann Lee in (1780), organized their micro-society in aversion to mercantile culture. They wrote little about the craft utopia they were building; their glorious round barns, their clothespins, their boxes and spindles and hooks endure, though their farming community, with its gender equality, does not. There were colonies of Free love Christians at Oneida, NY (1848) and Transcendentalists at Fruitlands, MA (1840). There was the morally righteous rural education scheme, Chautauqua (1874), and Seneca Falls meeting (1848) built the claims for women's suffrage. (Ok, thats not farming, but it was a reform movement, and it did succeed, and it started in upstate NY.)

   During and after the Civil war, with the dismantling of the slave-economy, a tremendous literary uproar rose up in protest against the industrial, mercantile north. Read The Velvet Horn, by Andrew Lyle, or the collection of essays I'll Take My Stand (1930) to find the generation of disinherited and unapologetic sons of the confederacy who mourned their short reign. While their version of golden agrariana was based on a plantation system, with slavery and cotton (for industry) as a major driver-- still there are parts of their narrative that are 'anti-mercantile' for moral reasons having to do with connection to place and tradition. Though we cannot forgive them their slavery, I feel that human and relational context that they examine is totally legitimate. By this time, in America things are happening fast! Our favorite farmer president Thomas Jefferson signed the deal on the Louisiana purchase, opening up the narratives possibility of The West, and opened a frontier full of adventurers. Lewis and Clark, Calamity Jane, speculations, pioneers, emigration and settlers there's books about all of them. Check out: My Antonia(1918), Little House on the Prairie,(1932) The Great Hunger(1962) for the classic stories of American immigrants. With the railroads, Capitalism came west, distorting land prices, inflating (then deflating) grain prices, centralizing and organizing the efforts of men. Nature's Metropolis (1991), The Jungle (1906), describe the conditions that gave birth to Populism and revolt, bringing on the labor movement. There were boomtowns, gold rushes, Klondikers, Sod busters, Lumbermen, Ranchers, Cowboys, Indian Wars galore, Its not easy to find the indigenous viewpoint on the taming of the continent. Though there are mourning stories, the westward narrative is as much about displacement as opportunity. I prefer Bones of Plenty(1962) to Grapes of Wrath,(1939) its about North Dakota wheat farming families during the beginning of the Great Depression.

The great depression preceded WW1, and with it the dawn of agro-chemicals, massive , and distortive export to feed armies and an inflection point for the course of agricultural economies and mechanics. Reading now for the reaction by farmers and thinkers to this big shift in farming methods, it is interesting to find again this theme of " lost time" we have yet to fully explain. Over in Germany, Rudolf Steiner's Lectures on Agriculture (1924) is a case in point. Steiner was asked to give these lectures by farmers who noticed a decline in animal and crop health within their lifetimes, just at the dawn of industrial method farming. They were concerned by what they observed, a "loss in vitality," which we today ascribe to declines in soil health, compaction, bad animal nutrition, and loss of micro-biodiversity as well as plant biodiversity from their farms. The farmers sought Steiner's advice on how to remediate the underlying " life forces" in agriculture. His lectures and the 'preparations' he described were a result of Steiner's study of " pre-Aristotelian" philosophy, what he called Anthroposophy. He preached a management based on observation of heat, light, cold, dark, metals, minerals, using these as signs and signposts to describe the underlying qualities of each element in the system.

Later in the almanac you'll read about the basic astrological/ lunar influences that Steiner advised for different activities. His guidance is still practiced widely. Every year Biodynamic farmers stuff skulls full of bark chips and mesenteries full of dandelions. they stuff cow horns with manure and intestines with chamomile flowers. They do it together, formally if not ritualistically. These preparations are highly biologically active and act as "inoculants" to the soil, and they are created with intention. Whatever your feelings about the micro-additions of these compounds, all it takes is a walk around a well run biodynamic farm to see that the care and observations of the farmer yields a strong system and very robust produce. This kind of farming requires a learned sensitivity to the complexity of agro-ecosystems. Steiner's concern was how to make that system sing with health, become introverted, cycling and recycling nutrients within the farm itself. The farmers' concern is improving the soil, importing less materials, and developing what Steiner would call its "Character," accelerating what my friend Dorn Cox (GreenstartNH.org) would call "biological velocity".

rough tongue clogs the obstructed jaws.

'Twas helpful through inverted horn to pour

Draughts of the wine-god down; sole way it seemed

To save the dying: soon this too proved their bane,

And, reinvigorate but with frenzy's fire,

Even at death's pinch-the gods some happier fate

Deal to the just, such madness to their foes-

Each with bared teeth his own limbs mangling tore.

See! as he smokes beneath the stubborn share,

The bull drops, vomiting foam-dabbled gore,

And heaves his latest groans. Sad goes the swain,

Unhooks the steer that mourns his fellow's fate,

And in mid labour leaves the plough-gear fast.

Nor tall wood's shadow, nor soft sward may stir

That heart's emotion, nor rock-channelled flood,

More pure than amber speeding to the plain:

But see! his flanks fail under him, his eyes

Are dulled with deadly torpor, and his neck

FOOTNOTES

You will find throughout this Almanac citations from primary materials which we hope drive your interest in studying these histories, particularly the parts of them that happened where you now farm. It is our goal for the 2014 Almanac to create an annotated bibliography ' open source reading list for an imaginary degree in An Alternative American Agricultural History. (AAAAH) This will be a collaboration, and we will need your suggestions. A new collection of agrarian writings deserves mention :American Georgics(2011) with interpreted excerpts from Creveceouer to Wendell Berry, thank you Yale University press. A Gem! This is a great starting point for the amateur historian, a sampler and "spine" to build from.

Sinks to the earth with
drooping weight. What
now

Besteads him toil or
service? to have turned

The heavy sod with
ploughshare? And yet
these

Ne'er knew the Massic
wine-god's baneful boon,

Nor twice replenished
banquets: but on leaves

They fare, and virgin
grasses, and their cups

Are crystal springs
and streams with running
tired,

Their healthful
slumbers never broke by
care.

Then only, say they,
through that country side

For Juno's rites were
cattle far to seek,

And ill-matched
buffaloes the chariots
drew

To their high fanes. So,
painfully with rakes

They grub the soil,
aye, with their very nails

Dig in the corn-seeds,
and with strained neck

O'er the high uplands
drag the creaking wains.

No wolf for ambush
pries about the pen,

Nor round the flock
prowls nightly; pain more
sharp

Subdues him: the shy
deer and fleet-foot stags

With hounds now
wander by the haunts
of men

Vast ocean's offspring,
and all tribes that swim,

On the shore's confine
the wave washes up,

Like shipwrecked
bodies: seals, unwonted
there,

Flee to the rivers.

Georgics book 3

Dorn is a student of the Physiocrats (1760s), a group of French proto-economists who understood national self-interest in terms of soil health and agricultural potential. Since the publication of Adam Smith's Wealth of Nations (1776), we have seen the focus of economics on wealth accrued through gold and conquest, industrial development and petro-chemical power. But just a few years earlier, the power of a state was directly correlated with the agricultural surplus, which translated into military power and capital for use in developing other functions (including sailing boats to the new world to capture gold). They realized and discussed the potential for land to yield, and increase in yield, over time via human management. Agricultural soil in particular, according the the Physiocrats is the key institution for long term prosperity. Frustrated by the artificiality of cities, of speculators, of mercantilists and conquistadors who distract from humanity's highest purpose, the Physiocrats worked to discover the logistics of optimizing our land use, of adding to the dynamism of the soil, and of orchestrating the "natural order" within our society, which would most closely match that of nature and allow humanity to co-exist with ecology. This inquiry into the nature of soil health is carried on by E.F. King in Farmers of Forty Centuries(1911), W. C. Lowdermilk's Conquest of the land through 7,000 years(1938), and more recently by Elaine Ingram in her Soil Food Web project. J.I. Rodale got us started started talking about "organic" farming in this country with Rodale Publishing(1930). He was inspired by the works of Sir Albert Howard, An Agricultural Testament (1940)who studied the manufacturing of compost in Indor, India, as a colonial agronomist. And he too made the connection to human health from the resultant produce.

Steiner saw the farm and the Waldorf schools as a source for health and reform that would ripple or shine out into all societies. The Enlightenment era Physiocrats similarly understood their work in terms of optimizing the health and power of the whole system. They believed that finding harmony with soil processes would be reflected all the way up the chain. All of these thinkers made the connection between the biochemical and socio-cultural systems that arise from soil management. They fall short of an indigenous or spiritual reverence, but all approach it, each in their own way espousing agrarian ways of being that naturally arise from contact with nature.

Getting back on track with our history lessons, we have to deal with the reactionaries of the 1930's and 1940's who set up the current organic movement. Ralph Borsodi's Flight from the City in 1933, along with Helen and Scott Nearing's Living the Good Life(1954) and Louis Bromfield's Pleasant Valley (1945), built on the growing notion of organic living, rejecting urban life, war, and mainstream consumerist values for philosophical and economic reasons. In the process they found a far more satisfying spiritual, literary and physical life in the country. In their own gardens and home-built homesteads they 'proved' the human and lifestyle values of a more rural, more principled existence. These were the intellectual godparents to the Back to land movement of the 1960/70s.

Their retreat into a self-made utopia gave them space to critique, and because they were self-sufficient, they were free to say it straight.

Part 2
A UTOPIA, IF YOU CAN BUILD IT

Farming, though it absorbs almost every moment in work, also gives an independence that is impossible for many other kinds of workers, in other words: liberty. Planting a single grain and harvesting it a hundredfold represents the resounding, emblematic possibility of human enterprise. In America, it means earning your own bread and voting for your own interests. The independent farmer, described by Thomas Jefferson" Cultivators of the earth are the most valuable citizens. They are the most vigorous, the most independent, the most virtuous, and they are tied to their country and wedded to its liberty and interests by the most lasting bands" This independence and virtue has persisted as a core element of the American character. Many of our country's most powerful leaders were also farmers, and they saw the tremendous metaphorical power of agriculture in the project of America.

What were the values first preached in our American Almanacks? They were practical, utilitarian, based on the ideas of the Enlightenment as interpreted (and simplified) by men like Benjamin Franklin who built anti-British sentiment with his "Poor Richard" moralism. Though some mercantile elites' sentiment ran with the colonizer, the ultimately victorious patriots were committed to a more egalitarian, more yeoman, more equal society. In this land full of farmers, it was Paul Revere, skilled craftsman, whose voice was heeded as he galloped at midnight to warn of the British invasion. He was well chosen in this task, for his trade afforded him credibility. We trust a man that works with his hands, to this day, which is why Republican candidates play country music at their rallies. Ben Franklin, like Merle Haggard,

brought a cultural atmosphere to the proceedings which shifted the terms of institutions in favor of ' the people' in a quite utopian way. The democracy they imagined at that time was truly revolutionary, given the context.

FC

These were the ideas of the Enlightenment, liberty, and social justice as interpreted by the designers of our democracy. "The earth, in its natural uncultivated state... was the common property of the human race." (Thomas Paine, Agrarian Justice) He went on to say that property owners were not inherently superior to the landless and that the poor ought to be provided for from the bounty of the more fortunate. " To the evil of monarchy we have added that of hereditary succession; and as the first is a degradation and lessening of ourselves, so the second, claimed as a matter of right, is an insult and imposition on posterity." (Thomas Paine, Common Sense) These convictions provided a useful motivating sentiment for innovation, industry and progress. All very fitting for a country built by opportunistic immigrants claiming a vast continent for their empire. Inspired in this vein, Jefferson wrote the gloriously magnanimous -- and some would say unnecessarily universalist -- treatise, The Declaration of Independence, to start us off right.

"We hold these truths to be self-evident, that all men are created equal, that they are endowed by their Creator with certain unalienable Rights, that among these are Life, Liberty and the pursuit of Happiness.--That to secure these rights, Governments are instituted among Men, deriving their just powers from the consent of the governed, --That whenever any Form of Government becomes destructive of these ends, it is the Right of the People to alter or to abolish it, and to institute new Government, laying its foundation on such principles and organizing its powers in such form, as to them shall seem most likely to effect their Safety and Happiness."

Safety and Happiness! The goal of independence for which these founders were willing to wager their own personal prosperity, was bound up in a fundamental commitment to equality. The proto-aristocracy of this fledgling colony had rejected the whole premise of an aristocracy, to their lasting credit.

"And for the support of this Declaration, with a firm reliance on the protection of divine Providence, we mutually pledge to each other our Lives, our Fortunes and our sacred Honor."

From his study, Jefferson could look out over his glorious vegetable gardens, and past them to the Blue Ridge Mountain range and westward on the valley below. He felt everyday the expanse of possibility ahead, and when his moment came, he signed the Louisiana Purchase to double the size of American and push westward.. From his office he would walk out on a long porch to meet with his farm supervisors, a porch whose foundation housed slave quarters, hot outdoor kitchens, and workshops where 7-year olds manufactured iron nails, keeping his farm financially afloat. Jefferson read Voltaire, Rousseau, Locke, and corresponded feverishly. Though he kept slaves, and ran an insolvent plantation, when it came to the knowledge of farming he adhered to his own principles. "He who receives an idea from me, receives instruction himself without lessening mine; as he who lights his taper [(candle)] at mine, receives light without darkening me."- Thomas Jefferson He meant by this a non-exclusive kind of practical knowledge (from economic botany,

to mechanical inventions), that should be shared that all may prosper by these developments. De Tocqueville, an early commentator on the American experience, identified this trait as one of the great triumphs of American Democracy: "self interest, rightly understood." Holding power and privilege in this newly sovereign nation was indeed a blessing, but America must certainly not replicate the feudalism and aristocratic patterns of Europe-- which being based on exploitation and imbalance are precarious. Jefferson said: "The price of freedom is eternal vigilance.-- Therefore for our own best long-term interests we must cherish reciprocity,mutual benefit, universal opportunity.

"That ideas should freely spread from one to another over the globe, for the moral and mutual instruction of man, and improvement of his condition, seems to have been peculiarly and benevolently designed by nature, when she made them, like fire, expansible over all space, without lessening their density in any point, and like the air in which we breathe, move, and have our physical being, incapable of confinement or exclusive appropriation." -Thomas Jefferson I read this to mean "open source." In those days it was via correspondence and printed journals instead of online forums, but in the same way it was for the betterment of mankind. In his way, Jefferson strove tirelessly to inform the many untrained farmers in this country in the ways of progressive farming, soil conservation, horticulture, hemp and fiber, cover crops and rotations. Indeed, many of them were poor, or even slovenly farmers. He imported and bred vegetable, fruit and grain varieties from across the world, particularly wine grapes and orchard trees, with

FOOTNOTES

From Thomas Jefferson's correspondance, downloaded off Archive.org, printed out by Severine and brought to Farm Hack website design meeting where it was posted by web-developer RJ to the first Farm Hack forum.

No man e'er was glorious, who was not laborious.
Would you live with ease? Do what you ought, not what you please.
He that cannot obey, cannot command.
An innocent Plowman is more worthy than a vicious Prince.
I never saw an oft-transplanted tree, nor yet an oft-removed family, that throve so well as those that settled be.
Reading makes a full Man, Meditation a profound Man, discourse a clear Man.

-Benjamin Franklin

all the fervor of a patrician generalist. Even as president he maintained an active horticultural correspondence.

"No sentiment is more acknowledged in the family of Agriculturalists than that the few who can afford it should incur the risk and expense of all new improvements, and give the benefit freely to the many of more restricted circumstances. ... I will throw out a first idea, to be modified or postponed to whatever you shall think better."

Franklin, Jefferson, Paine and many others demonstrated a line of thinking, a way of approaching progress that our current mainstream farm community seems to have almost forgotten. It is a relic trajectory of thought, now collected together in anthologies and nature magazines, or in used bookstores under the heading "Americana." It holds utopia like a carefully folded secret inside a rational approach to land surveying, cover cropping, the layout of farm buildings, the judicious development of waterways and small town values. I'm not forgetting the development agenda, the slavery, the commodification of nature, the subjugation of women. But there are certain values embedded in this era of history that we can build from, these principles that drove these men to design a social and economic system in alignment with natural order, to create a "science of liberty". And they aren't hidden or marginal, the are the founding principles, the founding values, the first words on the documents that describe our nation state. Thomas Jefferson would chide us for giving up

on such a noble structure, allowing it to become corporate, bloated and corrupt.

Part 3
UNDERSTANDING OURSELVES

Franklin wrote his 'truisms' inside the format his Poor Richard's Almanack, as the Almanac was at that time the correct format for communicating with farmers, as well as townspeople. It told the weather, the conversion tables, and it spread political messages to the makers of our new nation, our utopia. Is the almanac the right format for today? My opinion, as editor, is "yes," that farmers and farm culture are again the appropriate place for new ideas to wedge inside of.

I chalk it up to young muscles flush with blood and fresh air. Daily interaction with nature yields an almost unrealistic optimism. Farm startup is an undeniably utopian action, at once personally, and institutionally empowering. Today's young farmers, falling into a long lineage of young farmers, are system-builders with a sense of their collective impact. We live in the old economy, and we sell into it, but most of us are farming farms that hold different values, and the culture of farming is what gives us the strength to hold a perimeter around our microcosm, the generative power of life giving us the currency to persist. I'd argue that the seed for a radical transformation of in our economy can be found in farming and in the radical culture of farming as the resonant outcome of human agency, human values, and an accelerating dynamic that will shift the expectations of our children as well as the children of those that we are feeding with our produce.

As individuals and farmers, we govern the design of our farm systems and community inter-actions, and we live with the outcomes of those decisions. In being married to a place, we are uniquely accountable. Unlike the majority of non-manual, non-actual laborers in our society, we begin to recover the skill to observe which actions have what consequences. And we are forced to watch patiently, to adjust our management. The faith, in this case, is that our actions can yield a sustainable farm, a profitable farm, an improved community of relationships, that our actions can bring about greater prosperity for our children and the land. Growing our own capacity for practical knowledge -- what the ancient Greeks called " metis" -- that is the practical outlook that carried Odysseus through his whirlwind of trials in unknown parts. It is a human and lived practice that we are practicing.

The ambition we bring to agriculture enlivens the land we land on and community we land in. It also drives us to work on big questions.

CAN WE IMPORT CARBON FROM THE ATMOSPHERE?

CAN WE PRODUCE MINERAL-RICH, NUTRITIONALLY VITAL FOODS?

CAN WE HARMONIZE OUR FARM WITH THE LARGER ECONOMY, EVEN IF THE TERMS OF THAT ECONOMY ARE UNREASONABLE AND DISTORTED?

CAN WE SEE COMMERCE WITHIN A CSA AS SCAFFOLD ON WHICH TO REBUILD GENUINE COMMUNITY RELATIONS? CAN WE SUCCEED WITH NON-TRADITIONAL GENDER ROLES AND REGENERATIVE ROTATIONS?

CAN WE MIMIC THE ACTION OF PREDATORS ON OUR HERDS SO THAT THEIR HOOF-CHURNING WILL BUILD SOIL AND FIX CARBON?

CAN WE DO IT ALL, HAVE BABIES, RUN FOR TOWN COUNCIL, SERVE ON A NATIONAL BOARDS, PRESENT AT CONFERENCE, WRITE, READ, STUDY, AND ALSO HAVE ENOUGH STILLNESS FOR OUR SOULS?

CAN WE ACHIEVE PROSPERITY AND DIGNITY IN OUR DAILY WORK USING TECHNOLOGY THAT WE DRIVE FOR OURSELVES?

Forced to make thousands of decisions daily, as we wrestle with own farm designs, even alone and with best intentions, we unwittingly engage with the inherited values of traditional farmers: their placement of the barns, shelterbelts and windbreaks, their layout of pastures and drainage, the gender roles of killing, plucking and eviscer-ating, of bookkeeping and tillage. The mucky compaction layer where cows stood waiting to be milked, the poorly built farmworker or share-cropper housing, the eroding irrigation ditch, the pipe flushing manure into the nearest stream. We make some of the same decisions and some different ones. We inherit good and bad, toxic legacies and brilliant designs. We have heritage genetic code in our vegetables based on older people's tastes, storage needs, cider preferences. We learn their values through doing, and failing, and trying again. We ' pick it up' as we go along.

We may be amateurs, but farming is as good a way as any of gaining experience, and quick, even if we don't own the land. It requires ob-servation and manipulation of the ecosystem, of a living complex landscape. The significance of this practice is neither observable to the outsider nor subjected to routine commentary or analysis by society. The outcome issues from the actions of life and time: a micro-evolution of the species, the harmonizing or de-harmonizing of interact-ing elements, a generative process that is either working, or not. Feedback loops either register or don't. It is intimate. It is silent. It is direct and personal. We measure with our arms the weight of wood that burns right in front of us, creating a known unit of warmth. The bacon hangs in the smoke, then gives off more than just aroma, but also a feeling of security. We can feel it and know it.

Hut happiness is a protective state sought continually by some and occasionally by all as an escape from events and prospects which dismay. The human being is most sufficient when least in contact with conditions which dwarf his personality and suggest the insignificance of his tenure, his works and his emotions...Within garden walls his tenure, works and emotions have importance created by the nearness of his horizon...Thus one raises the walls of hut happiness and seeks an intensive life of small details made significant... The sap runs, Gardens are planted. Fields ripen. Harvest comes." -Clifford and John's ALMANACK 1680.

We straddle agrarian culture and market culture, to be practical. We sell to the markets. We barter between ourselves. Capital and property are problematic for many of us, often because we don't have much of either by the time we've come through college and have put in enough study-hours to understand the basics of our complex world. As a generation with 1 trillion dollars in educational debt, we are overeducated for this economy. Not only have our colleges failed to teach us practical skills, but we've been digitally acculturated into expecting instant, predictable information, wifi access, seamless interfaces -- a kind of techno-informational old people's village. To start a small business in this economy is to discover very quickly that the FSA is not Dropbox, and that we'll have to relearn how to do things the old fashioned way, with dial up, secretaries, red tape, bigotry and fax machines. The human logistics of startup means overcoming these hurdles, playing the ' game level' to get what we need, wagering on our own capacity to make the impossible happen, over and over again.

Part 4
WHAT WE BELIEVE IN GUIDES US

Pious Superstitions

Though superstition, old and stale,
Or strange beliefs in false or true,
Mar, rather than adorn the tale,.
they, needs, must be recorded too;
For in them lie much of the lore
And legend of th' unwritten page
Of years gone by, three score or more,
When Faith, not reason, sat as Sage.

But true or false, or whence they came,
We little know and care still less;
Our sires believed them, all the same,
And to believe was to be blest;
The faith-cure all assaults withstood-
A double virtue had each charm-
costless, and if it did no good,
It certainly could do no harm.

-H.L.Fisher

York, Penna. 1888

Though we are nostalgic, with the internet at our fingertips we aren't hemmed in by an "either or." We do not rely on the Almanack anymore for our insights on the weather, nor are we superstitious. We love to gossip with our old-timer neighbors, but we are not satisfied to farm in their way. The sector is riddled with complex tensions that includes concessions and work-arounds, hypocrisies and outdated dogmas. Some of us farm with horses, but map our fields with GIS tools, move our animals with temporary fences controlled by Arduinos. On either side of our farm we're likely to find social conservatism paired with biotechnology. Our neighbors may hold tremendous wealth in land but suffer from intense stress, financial precarity, byzantine regulations, unreliable worker programs, disinterested heirs. Wouldn't it be simpler if we could talk only about a standoff between rationality and superstition, between commodity and conviction, between conventional and organic? But we cannot -- our landscape matrix is far more muddled. It is to be expected that this practice which straddles nature and industry would be complicated.

We are new to these theories. We are amateur practitioners of agrarian culture. Since many of us were born in cities and suburbs, we're slowly acculturating ourselves to agrarian ways predicated more on mutuality than competition, mostly by reading Wendell Berry, learning to enjoy a quiet sit on the front porch, and finally having space enough to become more humble versions of ourselves. As amateurs, we have much to learn by studying history for precedents as we strategize the next big moves. We know that sharing is a good business strategy, and there is strong interest in forming more cooperative feed shops, in cooperative farm purchase, in microfinance, crowdsource-funding streams, and open source technology sharing. There are innovators within our movement who are creating institutions that shift behaviors and practices. FOOTNOTE OurGoods.org co-founder Caroline Woolard recently spoke at our Farm Hack www.farmhack. net The talk she gave is included in the audio podcast that accompanies this almanac.) event in New York City. She talked about ' learning to share' and having to face up to a new set of practices, instincts and behaviors of mutual aid. We mustn't cling to the market value of our work; we mustn't haggle for the maximum return. Instead, we are building out a personal network of collaboration that may well extend beyond the current project, the current transaction. It's a kind of complex reciprocity that takes practice, even in small towns, let alone in the competitive creative community of New York City. She urged us to practice these skills that we may benefit by them. See also: Sheepscot General Store http://www.

sheepscotgeneral.com/

Wendell Berry says: "Change comes from the people at the bottom doing things differently." It's a theory of change that seems foreign in a world of TED talks and mesh-network technologies, but if we hold to these foundational theories of the system on which this nation is built, and hold close to our hearts the fact that more than 70% of the world's population is fed by small farmers like ourselves, it's easier not to be confused by all the media hoopla.

Remind yourself of some truths: All flesh is grass (Isaiah 40:6). All life depends on photosynthesis and on a finite layer of soil. There is a limit to growth, a limit to wealth, a limit to everything. You can't eat money. Pop-culture has forgotten a lot of things Don't bow to its logic.

The pickle we are in is certainly man-made, and it may be a pickle far older even than the best of farming. But it does seem like there is something we can find, in farming, that can shift both our own lives and that of the larger culture in a better, stabler direction. There is a personal kind of warmth, prosperity, and comfort, which allows us to recapture some of that trickster, the survivor, the out-witter. Its vague and big, but its a kind of general Homo Sapiens' mojo that seems to always have been an undercurrent in our culture: the group of crazies who have been behind many social innovations and resistance movements, the Benedictine monks, the underground railroad, cave-dwelling mathematicians, unglamourous struggles against injustice throughout time, the persistent undercurrent of humanity that dreams of universalism, morality, utopia, freedom.

FOOTNOTES

Clifford and John's Almanack 1816

"When we grasp fully that the best expressions of our humanity were not invented by civilization but by cultures that preceded it, that the natural world is not only a set of constraints but of contexts within which we can more fully realize our dreams, we will be on the way to a long overdue reconciliation between opposites which are of our own making." - Paul Shepard Coming Home to the Pleistocene (1998)

The genetic momentum we're working with, the wits we employ, our problem-solving and hustle, these may well be older skills than farming, but they are strengthened by farming. But farming is a way to bring our minds and our bodies into a state where we can feel more freely, because we become accustomed to challenges -- daily, persistent, never ending challenges. And we'd certainly have more practice at overcoming challenges, having put in the time to learn stress management. A personal economy, and a personal practice up against nature as well as the market, puts us in a far better position to evaluate the choices we face in authoring a livable future. "Warmth, earned by effort, arises within, and the odor of cakes makes the late rising of the sun aromatic. ...A day of pleasant routine with labor made dignified by duty, the solvent of all human difficulties and distresses, the matchless organizer of humanity."

Whichever way we go, let's do it with purpose, and rosy cheeks.

- Severine v T Fleming, editor
The Greenhorns, director
Smithereen Farm, Essex NY

Thanks go to Audrey Berman, Tess Diamond, Francesca Capone, Drew Heffron, Alanna Rose, Fay Strongin, Garth Brown and Ines Chapela who prepared, scanned, edited, copy-edited, corresponded and photoshopped to make this volume happen. To the hundred essayists, contributors and illustrators who sent in the articles contained. To Rick and Megan Prelinger, who encouraged me to peep into the archive, pulled out an old box of frayed Almanacs and patiently coaxed me to understand the community power of history. What a blessing, those dozens of days I spent zig-zagging in their stacks, freely appropriating, figuring out the order of events at my own punky pace. Thanks also to Anna Duhon and Conrad Vispo of the Farmscape Ecology program, who shared many titles on Agricultural history in the Hudson Valley and inspired the creation with Markley Boyer of the 'Changing Hudson River Landscape' poster. A nod to my parents Ronald Lee Fleming and Renata von Tscharner of the Townscape Institute, Cambridge MA, whose 'New Providence' poster series we mimicked. To Amy Franceschini and Sledge Taylor who patiently dredged up articles and images from their own strong archival practice, who collect stories themselves and showed me how to search. To Tom DePietro, Chris Carlsson, Matthew Stadler at Publication Studio, Patrick Kiley at the New York Public Library for guidance on DIY publishing. To Kristin Kimball, for setting up "writers mornings" where we typed furiously together, in silence. To Daniel Bowman Simon, for his many dog-eared printouts, and Henry Tarmy for revisiting The Land newspaper archive of the 1940's, to CJ Sentell who tackles the winding root of slavery in Agriculture- their powerful practices inspired me to be braver. To Kenneth Mroczek for donating to us the Kelsey letterpress we used to print the cover, the size of which determined our book, and to Alessandro d'Ansembourg and the Flora Foundation, who trusted our vision and supported the printing of this first 2013 New Farmer's Almanac. We look forward to working with yet more of you.

Due date for submissions to the next Almanac : June 2013.
Contact: almanac@thegreenhorns.net

OTHER PROJECTS BY THE GREENHORNS
Greenhorns: 50 Dispatches From The New Farmer's Movement, *Storey, 2012*
A Fieldguide For Young Farmers, *For Highschoolers And Educators*
Punk Yeoman, *A Greenhorns Guide To Organizing*
Affording Our Land, *A Greenhorns Guide To Finance*
Land. Liberty. Sunshine. Stamina, *A Compendium On Land Access Issues 2010*
The Greenhorns, *A Documentary About The Spirit,*
Practices And Needs of Young Farmers In America
Ourland, *A Web-Series About Building A New Food Economy Inside The Old One*
Greenhorns Guide For Beginning Farmers

JOIN THE NETWORK
thegreenhorns.net
thegreenhorns.wordpress.com
serveyourcountryfood.net
farmhack.net
ourland.tv
youngfarmers.org
heritageradionetwork.com
newfarmersalmanac.org

EDITOR IN CHIEF & PUBLISHER
Severine von Tscharner Fleming

MANAGING EDITOR
Audrey Berman

DESIGNER & ART EDITOR
Francesca Capone

EDITOR-AT-LARGE
Garth Brown

EDITORIAL TEAM
Christopher Chemsak
Tess Diamond
Louella Hill
Lulu McLellan

DESIGN DIRECTION
Drew Heffron

CREATIVE CONSULTANT
Amy Franceschini

COVER ILLUSTRATION
Alanna Rose

INDEX OF AUTHORS

Kristen Kimball

Latta, Zoe
New York, NY
zoelatta.com

Lamson, Mariette Brooklyn, NY

Lehrer, Ava Queens, NY
Assistant to the Director of
the Unterberg Poetry Center
at the 92th Street Y

Loria, Kristen Hudson, NY
kristen.loria@thegreenhorns.net

Oeding, Schirin Rachel
Southern Ontario
Farmer & Handcrafter

OurGoods
New York, NY ourgoods.org

Paper Tiger Video Collective
New York, NY
Filmmakers / Media Collective
papertiger.org

Millonzi, Katharine
Ecological Gastronome
neweconomicsinstitute.org

Elise McMahon
Hudson, NY
Furniture Maker, Community
Organizer, Educator
elisemcmahonmade.org

Moschovakis, Anna
South Kortright, NY
Writer/Editor/Translator
badutopian.com

Prelinger, Megan
San Francisco, CA
Library-Builder, Writer, Wildlife
Rehabilitator. prelingerlibrary.org

Prelinger, Rick
San Francisco, CA
Archivist, Filmmaker, Librarian
prelinger.com

Anastasia Rabin
Tuscon, AZ

Raissian, Katie
Brooklyn, NY
Editor-in-Chief of
STONECUTTER: A Journal
of Art and Literature
stonecutterjournal.com

Ramsey, Trace
Durham, NC
cricketbread.com

Roman-Alcalá, Antonio
San Francisco, CA Urban Farming
Educator, Artist, Musician, and
Human Enthusiast
antonioromanalcala.com

Schiffler, Kat
Lyons, NE
Storyteller, Center for Rural Affairs
cfra.org

Schober, Deborah

Skinner, Jonathan
Bowdoinham, ME ecopoetics.org

Snyder, Sarajane
White Deer Township, PA
fairweatherly.com

Tarmy, Henry

Trzaskos, Brian
Essex, NY Physical Therapist
ascentwellness.com

Waldrop, Rosmarie
Providence, RI
Writer, Educator, Co-Founder of
Burning Deck Press

Willey, Tom

Zoninsein, Leonora
NYC, Rio de Janeiro Civic
Participation Strategist, Meu Rio
cargocollective.com/upfront

INDEX OF ARTISTS

Barnet, Becca
Charleston, SC
Artist beccabar.net

Black, Hannah
Brooklyn, NY
hnnhsblck@gmail.com

Blankenship, Jana
Accord, NY
Scent Artist, Beauty Maker, Curator
captainblankenship.com

Beery, Josef
Free Union, VA
Artist, Printer
josefbeery.com

Beggs, Michael
Bethany, CT

Boyer, Markley
Dudson Valley, NY
Multimaven

Budner, Brooke
San Francisco, CA
Farmer, Artist

Calisch, Nolan
Portland, OR
Farmer, Artist

Capone, Francesca
New York City
Designer, Artist, Writer, Educator
francescacapone.com

Cline, Laura

Cowgill, Jacob
Power, Montana Farmer
PrairieHeritageFarm.com

Engleman, Lucy
Grand Rapids, MI
Freelance Illustrator
lucyengelman.com

Franceschini, Amy
San Francisco, CA Artist
futurefarmers.com

Galloway - Kane, Ceilidh
East Hardwick, VT
ceilidh.g.kane@gmail.com

Gipe, Shayne
Westchester County, NY
Writer, Moonlighter, Farmer-in-
Training
thecotyledon.com

Howard, Ashley E.
Texas Artist, Photographer
ashleyehowardstudio.com

Horstman, Fritz
Bethany, CT
fritzhorstman.com

Huggins, Adam T
Turtle Island, NC
Permaculturalist, Two-Wheeled
Propagator Of Plants and
Emulsions
sunfishmoonlight.wordpress.com

Jenkins, Trish
Spearfish, SD
Farmer
cyclefarm.wordpress.com

MacLean, Alex
Lincoln, MA
alexmaclean.com

Maki, Ginny
Artist and Server
ginnymaki.com

Malaskey, Stacey
Detroit, MI
Beekeeper, Artist, Printmaker
staceymalasky.com

Maresca, Lauren
Brooklyn, NY Artist
laurenmaresca.com

Montenegro, Sonya Elena
Chicago & Fairbanks
Artist & Aspiring Farmer
sonyamontenegro.com
etsy.com/shops/
BlackMountainGirls

Mroczek, Kenneth

Muller, Allison
Kansas City, MO
Designer
akmkiosk.squarespace.com

Norzi, Mateo

Pellizzi, Aurora
New York aurorapellizzi.com

Prefer, Marisa
Here Freelance Collaboration
marisaprefer.com

Reitman, John-Elio

Ripley, Christin
New York, NY
christinripley@gmail.com

Rose, Alanna
Central, NY
Artist/Farmer
alannarose.com, cairncrestfarm.
blogspot.com

Schumann, Peter and Elka
Glover VT Founders of Bread and
Puppet Theatre

Smith, Jeremy
Spearfish, SD Farmers
cyclefarm.wordpress.com

Jocelyn Spaar
Greenpoint, Brooklyn
Artist, Writer, Translator

Tareila, Emily
San Francisco, CA
Artist, Facilitator
emilytareila.com

Taussig, Olivia
New York City
Animator, Artist
olivia-taussig.tumblr.com/

Wali-Richardson, Kanchan
Brooklyn, NY Artist
kanchanrichardson.com

Willow, Gabriel
Brooklyn, NY
Educator, Biologist, Illustrator

Woolmington, Iona
Burlington, VT
Farmer, Artist at Iona Fox Comics
ionafoxcomics.com, facebook/
ionafoxcomics

Young, Christie
Brooklyn, NY Illustrator
pleasurezone.me

Young-Ramsey, Adrienne
Nelson County, VA Homesteader,
Mama, Musician backyardrevolu-
tion.com, adrienneyoung.com

Our Printers

SMALL OAK PRESS AND BINDERY

located near Hudson, New York, creates beautiful objects using traditional hand-bookbinding and print-making techniques. The studio produces a variety of bound and printed goods including journals, sketchbooks and albums, custom boxes and portfolios, as well as letterpress cards and paper goods. Find Small Oak Press and Bindery online at www.etsy.com/shop/smalloakpress or email at smalloakpress@gmail.com.

HAMILTON PRINTING COMPANY

is a 98 year old, family owned business that specializes in single color hardcover and paperback books. Hamilton Printing opened in 1912 as a small letterpress shop that focused on pamphlet binding. Through the decades, we increasingly found business opportunities in nearby New York City as publishing companies solicited us for their typesetting, printing, and paper needs. The resulting success allowed us to invest in additional presses and, in 1966, a paperback bindery. A decade later we purchased a hardcover binder and, for the first time, were able to offer complete printing and bindery service to our customers.

Table of Contents

"The weather is unpredictable,
still we must stay organized."
-SvTF

Through the Glacier Recession
And the Arctic Slosh & Trade
We handed beasts masking tape
Kept lips un-quivered

–Mariette Lamson

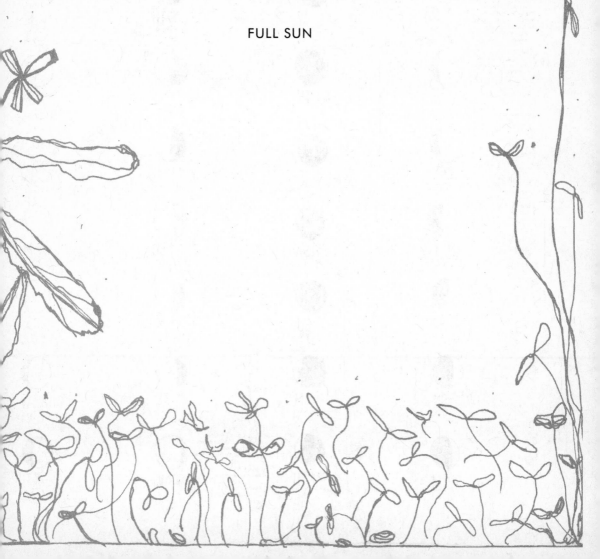

JANUARY

FULL SUN

JANUARY 2013

zodiac signs

Cancer	Scorpio	Pisces
Most Fruitful	Fruitful	Good for root crops but a very poor time to make seedbeds
Taurus	Capricorn	Libra
Semi-fruitful	Semi-fruitful	Time for setting of hay, grain, flowers and trees when barely is intended
Aries	Sagittarius	Aquarius
Semi-barren	Semi-barren	Barren
Virgo	Gemini	Leo
Barren	Barren	Most barren sign

1 In 1863, any U.S. citizen (male or female) who had never taken up arms could claim up to 160 acres and take title by living and farming on the land for five years for $18.
The Quadrantid Meteor Shower, which radiates from Bootes, peaks tonight through the 5th. It can best be seen after midnight.

2 1839: The first photo of the Moon thought to be recorded by photographer Louis Daguerre

It was an old Saxon belief that 2nd January was one of the unluckiest days of the whole year.

3

4 Martyrs of Colonial Repression Day

5

6

7

8 When two planets are 180 degrees apart from each other in the zodiac, it is called an opposition. On this day, the moon and Jupiter are in opposition.

9

10

11 NEW

12

13

14

15

16 Revolution Day in Tunisia

17

18

19

20 When two planets are 180 degrees apart from each other in the zodiac, it is called an opposition. On this day, the moon and Jupiter are in opposition.

21 The BBC in London made its first world broadcast in 1930

22

23 2013: EcoFarm Conference Begins in Pacific Grove, California (eco-farm.org)
Southern SAWG Conference Begins in Little Rock, Arkansas: Practical Tools and Solutions for Sustaining Family Farms (ssawg.org)

24

25

26

27 FULL

28

29

30

31

A Guide to the Almanac Calendar

compiled by MARISA PREFER, KRISTEN LORIA & CHRISTOPHER CHEMSEK
moon + zodiac drawings by EMMA HARDEN

Here are some keys and diagrams in reference to the almanac calendar. The information included is made up of facts and figures that we thought might be relevant, exciting or useful for individuals working on, off and with the land, but is by no means a comprehensive guide to Biodynamic farming practices. Much of the planting, moon phase and zodiac data has been adapted

with the permission from the 2013 Stella Natura Biodynamic Planting Calendar and from Jack R. Pyle and Taylor Reese by way of their book Raising With the Moon: The Complete Guide to Gardening—and Living—by the Signs of the Moon (both of which are highly reccomended reads if you are looking to learn more about this kind of farming).

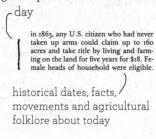

day

in 1863, any U.S. citizen who had never taken up arms could claim up to 160 acres and take title by living and farming on the land for five years for $18. Female heads of household were eligible.

historical dates, facts, movements and agricultural folklore about today

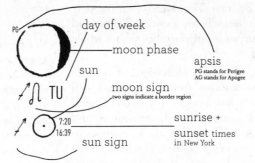

day of week
moon phase
apsis
PG stands for Perigee
AG stands for Apogee
sun
moon sign
two signs indicate a border region
sun sign
sunrise +
sunset times
in New York

MOON PHASE PLANTING

adapted from *Raising With the Moon*

All things planted in the First Quarter will seed outside the fruit. The Second Quarter is best for things that seed inside the fruit. The Third Quarter is for things that produce underground, or that depend heavily on an underground root system. The Fourth Quarter is recommended for planting only when no other time is available/possible. This is only a partial listing of vegetables. If you understand the theory, then you can determine the best Moon phase for any vegetable or fruit.

1st quarter	2nd quarter	3rd quarter	4th quarter
All grains	Beans	Beets	It is not recommended that you use the Fourth Quarter for planting. If dire circumstances arise, this entire quarter can be used for underground crops but will not be as productive as the Third Quarter.
Asparagus	Eggplant	Carrots	
Broccoli	Melons	Chicory	
Brussels Sprouts	Peas	Potatoes	
Cabbage	Peppers	Radishes	
Cauliflower	Pumpkin	Rutabagas	
Celery	Squash	Onions	
Corn	Tomatoes	Turnips	
Cucumbers	Flowers	Tubers	

ZODIAC SIGNS OF THE CALENDAR

Cancer Most Fruitful	Good for planting, irrigating, grafting + transplanting	**Scorpio** Fruitful	Exceptional for vine-type growth	**Pisces** Fruitful	Good for root crops but a very poor time to make sauerkraut
Taurus Semi-fruitful	A time to plant lettuce, cabbage, sturdy stalks + root crops	**Capricorn** Semi-fruitful	Good time to plant tubers + root crops	**Libra** Semi-fruitful	Time for seeding of hay, grain, flowers and crops where beauty is concerned
Aries Semi-barren	Time for cultivating + plowing where weed destruction is sought-after. Chickens hatched now will be patient	**Sagittarius** Semi-barren	Best for destroying unwanted plant life	**Aquarius** Barren	Suitable for weeding + destroying of unwanted growth.
Virgo	Good for bringing order to the garden + general cleanup.	**Gemini**	Great for cultivating + destroying unwanted growth.	**Leo**	A time for killing weeds + unwanted growth. Cider should be drawn off for vinegar • A good time to cut hair - if curly hair is desired, cut hair in the increase of the moon.

first photo of the moon, 1851 by John A. Whipple

zodiac chart from the citizen and farmers almanac, 1802

excerpt from Planting By the Signs: Mountain Gardening: The Foxfire Americana Library

RESOURCES

old almanac data: archive.org
events + facts: attra.org
biodynamics.com
anarchofacts:recollectionbooks.com
rudolph steiner: rsarchive.org
zodiac charts: stellanatura.com
tide charts: tbone.biol.sc.edu
early land movement:wholeearth.com

Almanacs In A Utopian Age

by RICK PRELINGER

Almanacs linked farmers with the work of earth and sky and the world of ideas. They thrived in an age of isolation that is almost unimaginable today, a time when there was no radio and newspapers moved slowly by mail. They were often the only books in their households besides Bibles, and they were meant to supply families with a whole year of reading.

Today most of us are a message or a flight away from one another (though we'd do well not to count on that forever). Our shared mind is clogged with rants, overnight sensations, and big ideas. More than ever we need almanacs to provide what we can't get online: carefully edited collections of smaller notions, hints, hacks and hard information that might appear simple, but when taken with a tall glass of water expand into mind-changing, load-lightening, actionable ideas. An almanac is a little book hiding an encyclopedia within its covers. Its job is to offer proverbs that turn into projects, household hints that help harvests flourish, facts that keep animals healthy and plants straight on their stems.

I love the puzzles in old almanacs, but I love even more how they conclude: "Solution in next year's Almanac." Patience is civil disobedience in our era of speed. Some things take their own time.

Rick Prelinger is the co-founder of Prelinger Library, an open and appropriation friendly private library in San Francisco. He is also founder of archive.org, the best place online to find free, open-source, out of copyright archival films, audio, scanned historic books and a treasure trove of passionate history projects. We begin here, and continue throughout the almanac to make notations about sources, pockets of treasures, lines of inquiry that you, reader, may care to explore for yourself.

EDITORIAL NOTE:

The Hudson River, home to Pete Seeger's Clearwater Sloop, is a heavily polluted corridor of commerce, and most famously home to a school of romantic painters of the mid-19th century. It was navigated and named for Henry Hudson, the explorer. It is a watershed as well as a foodshed to the urban island of Manhattan. As early as 1700's the banks of the Hudson were kept as hay meadows to supply the horses of New York City. Its port town of Poughkeepsie was formerly home to Delaval Dairy Processing, which supplied the heavy dairy trade up and down the valley. As dairies struggled an effort was made to convert to apples into cider, and a hard cider tradition persists to this day. The town of Hudson, 120 miles up river, was founded as a Whaling town by leaving Nantucket and Salem for calmer waters. It is hard to imagine massive 50 ton whales being towed by schooners up river for processing, or the smell of blubber being rendered. But that is history. Hudson has become a haven for antiquers, weekenders, artists, hipsters, and many families living in shambled subsidized apartments due to corrupt landlords and housing authority. - SvTF

TIDE CHART FOR HUDSON, NY AT THE HUDSON RIVER *by* MARISA PREFER

These tide predictions are from the University of South Carolina's tide predictor at tbone.biol.sc.edu. the values are for the Hudson River above the George Washington bridge and are based upon averages for the six months May to October, when the freshwater discharge is at a minimum.

january 2013

Day	High	Low	High	Low	High
Tue 01	05:18 3.52 ft	11:30 0.14 ft	17:02 3.99 ft		
Wed 02		00:09 -0.05 ft	05:58 3.54 ft	12:07 0.24 ft	17:45 3.89 ft
Thu 03		00:44 0.03 ft	06:40 3.62 ft	12:53 0.35 ft	18:34 3.78 ft
Fri 04		01:26 0.10 ft	07:27 3.73 ft	13:58 0.45 ft	19:28 3.67 ft
Sat 05		02:27 0.15 ft	08:19 3.86 ft	15:18 0.43 ft	20:27 3.57 ft
Sun 06		03:38 0.11 ft	09:16 4.02 ft	16:29 0.26 ft	21:32 3.51 ft
Mon 07		04:42 -0.03 ft	10:20 4.20 ft	17:31 -0.01 ft	22:45 3.54 ft
Tue 08		05:41 -0.22 ft	11:28 4.43 ft	18:29 -0.30 ft	23:58 3.68 ft
Wed 09		06:38 -0.44 ft	12:32 4.70 ft	19:25 -0.59 ft	
Thu 10	01:01 3.90 ft	07:34 -0.63 ft	13:29 4.93 ft	20:19 -0.83 ft	
Fri 11	01:58 4.12 ft	08:29 -0.79 ft	14:21 5.07 ft	21:11 -1.01 ft	
Sat 12	02:51 4.27 ft	09:22 -0.87 ft	15:12 5.08 ft	22:01 -1.09 ft	
Sun 13	03:43 4.33 ft	10:14 -0.85 ft	16:03 4.96 ft	22:48 -1.06 ft	
Mon 14	04:36 4.31 ft	11:03 -0.71 ft	16:56 4.74 ft	23:35 -0.90 ft	
Tue 15	05:30 4.23 ft	11:53 -0.47 ft	17:48 4.46 ft		
Wed 16		00:21 -0.65 ft	06:24 4.12 ft	12:44 -0.15 ft	18:40 4.14 ft
Thu 17		01:09 -0.34 ft	07:15 3.99 ft	13:39 0.16 ft	19:31 3.83 ft
Fri 18		02:00 -0.04 ft	08:05 3.85 ft	14:40 0.41 ft	20:22 3.54 ft
Sat 19		02:55 0.22 ft	08:54 3.72 ft	15:43 0.53 ft	21:14 3.31 ft
Sun 20		03:51 0.37 ft	09:47 3.63 ft	16:41 0.53 ft	22:11 3.16 ft
Mon 21		04:44 0.42 ft	10:42 3.60 ft	17:34 0.45 ft	23:11 3.13 ft
Tue 22		05:34 0.39 ft	11:39 3.67 ft	18:23 0.33 ft	
Wed 23	00:08 3.20 ft	06:22 0.31 ft	12:31 3.80 ft	19:10 0.18 ft	
Thu 24	00:58 3.34 ft	07:08 0.20 ft	13:16 3.95 ft	19:54 0.02 ft	
Fri 25	01:43 3.49 ft	07:52 0.07 ft	13:55 4.09 ft	20:37 -0.12 ft	
Sat 26	02:23 3.61 ft	08:36 -0.05 ft	14:31 4.19 ft	21:18 -0.25 ft	
Sun 27	03:00 3.71 ft	09:18 -0.15 ft	15:03 4.23 ft	21:56 -0.33 ft	
Mon 28	03:35 3.77 ft	09:58 -0.21 ft	15:34 4.23 ft	22:33 -0.37 ft	
Tue 29	04:09 3.82 ft	10:37 -0.22 ft	16:06 4.18 ft	23:08 -0.36 ft	
Wed 30	04:44 3.87 ft	11:15 -0.18 ft	16:42 4.10 ft	23:42 -0.30 ft	
Thu 31	05:23 3.92 ft	11:55 -0.09 ft	17:25 3.97 ft		

february 2013

Day	High	Low	High	Low	High
Fri 01		00:16 -0.21 ft	06:08 3.97 ft	12:40 0.05 ft	18:16 3.83 ft
Sat 02		00:57 -0.08 ft	06:57 4.02 ft	13:40 0.20 ft	19:11 3.68 ft
Sun 03		01:53 0.06 ft	07:52 4.06 ft	14:56 0.28 ft	20:11 3.55 ft
Mon 04		03:09 0.14 ft	08:52 4.10 ft	16:09 0.21 ft	21:18 3.46 ft
Tue 05		04:22 0.08 ft	09:59 4.16 ft	17:14 0.01 ft	22:33 3.48 ft
Wed 06		05:25 -0.09 ft	11:12 4.30 ft	18:12 -0.24 ft	23:47 3.65 ft
Thu 07		06:25 -0.31 ft	12:20 4.51 ft	19:08 -0.50 ft	
Fri 08	00:52 3.91 ft	07:21 -0.52 ft	13:18 4.73 ft	20:01 -0.72 ft	
Sat 09	01:47 4.18 ft	08:15 -0.69 ft	14:10 4.88 ft	20:51 -0.89 ft	
Sun 10	02:37 4.38 ft	09:07 -0.79 ft	14:58 4.90 ft	21:39 -0.97 ft	
Mon 11	03:25 4.48 ft	09:57 -0.79 ft	15:45 4.81 ft	22:24 -0.93 ft	
Tue 12	04:13 4.48 ft	10:44 -0.69 ft	16:33 4.61 ft	23:07 -0.78 ft	
Wed 13	05:01 4.39 ft	11:30 -0.47 ft	17:21 4.34 ft	23:49 -0.52 ft	
Thu 14	05:49 4.24 ft	12:17 -0.18 ft	18:10 4.05 ft		
Fri 15		00:31 -0.20 ft	06:36 4.06 ft	13:05 0.14 ft	18:59 3.75 ft
Sat 16		01:15 0.14 ft	07:23 3.87 ft	13:59 0.44 ft	19:48 3.49 ft
Sun 17		02:04 0.45 ft	08:10 3.70 ft	15:01 0.64 ft	20:39 3.27 ft
Mon 18		03:03 0.66 ft	08:59 3.56 ft	16:02 0.71 ft	21:33 3.13 ft
Tue 19		04:03 0.74 ft	09:54 3.48 ft	16:59 0.66 ft	22:33 3.10 ft
Wed 20		04:59 0.70 ft	10:55 3.50 ft	17:50 0.54 ft	23:33 3.18 ft
Thu 21		05:50 0.56 ft	11:54 3.63 ft	18:37 0.37 ft	
Fri 22	00:27 3.36 ft	06:38 0.39 ft	12:44 3.82 ft	19:21 0.19 ft	
Sat 23	01:13 3.58 ft	07:25 0.19 ft	13:26 4.01 ft	20:04 -0.00 ft	
Sun 24	01:54 3.80 ft	08:10 -0.01 ft	14:03 4.17 ft	20:45 -0.18 ft	
Mon 25	02:30 4.00 ft	08:54 -0.19 ft	14:37 4.29 ft	21:25 -0.32 ft	
Tue 26	03:03 4.17 ft	09:36 -0.32 ft	15:10 4.34 ft	22:03 -0.40 ft	
Wed 27	03:37 4.30 ft	10:18 -0.40 ft	15:45 4.32 ft	22:40 -0.43 ft	
Thu 28	04:13 4.39 ft	11:00 -0.40 ft	16:25 4.24 ft	23:16 -0.38 ft	

march 2013

Day	High	Low	High	Low	High
Fri 01	04:55 4.43 ft	11:43 -0.31 ft	17:12 4.11 ft	23:55 -0.26 ft	
Sat 02	05:42 4.42 ft	12:30 -0.15 ft	18:05 3.94 ft		
Sun 03		00:39 -0.08 ft	06:36 4.37 ft	13:29 0.05 ft	19:04 3.78 ft
Mon 04		01:37 0.13 ft	07:35 4.30 ft	14:39 0.20 ft	20:06 3.66 ft
Tue 05		02:53 0.28 ft	08:38 4.23 ft	15:51 0.20 ft	21:14 3.59 ft
Wed 06		04:07 0.26 ft	09:47 4.20 ft	16:56 0.07 ft	22:27 3.64 ft
Thu 07		05:13 0.11 ft	10:59 4.25 ft	17:54 -0.13 ft	23:38 3.84 ft
Fri 08		06:12 -0.10 ft	12:07 4.41 ft	18:48 -0.33 ft	
Sat 09	00:40 4.11 ft	07:08 -0.30 ft	13:04 4.58 ft	19:39 -0.51 ft	
Sun 10	01:33 4.39 ft	09:01 -0.47 ft	14:54 4.70 ft	21:28 -0.63 ft	
Mon 11	03:20 4.59 ft	09:51 -0.57 ft	15:40 4.72 ft	22:14 -0.67 ft	
Tue 12	04:04 4.69 ft	10:39 -0.59 ft	16:24 4.64 ft	22:57 -0.61 ft	
Wed 13	04:46 4.68 ft	11:24 -0.51 ft	17:08 4.47 ft	23:38 -0.45 ft	
Thu 14	05:29 4.58 ft	12:07 -0.34 ft	17:53 4.25 ft		
Fri 15		00:17 -0.22 ft	06:11 4.41 ft	12:50 -0.10 ft	18:40 4.00 ft
Sat 16		00:55 0.08 ft	06:55 4.20 ft	13:33 0.19 ft	19:28 3.76 ft
Sun 17		01:33 0.39 ft	07:39 3.99 ft	14:21 0.48 ft	20:17 3.54 ft
Mon 18		02:14 0.68 ft	08:24 3.80 ft	15:16 0.72 ft	21:06 3.37 ft
Tue 19		03:09 0.92 ft	09:11 3.64 ft	16:18 0.85 ft	21:58 3.26 ft
Wed 20		04:17 1.03 ft	10:02 3.54 ft	17:18 0.85 ft	22:54 3.24 ft
Thu 21		05:21 0.99 ft	11:00 3.51 ft	18:11 0.74 ft	23:53 3.32 ft
Fri 22		06:16 0.83 ft	12:03 3.59 ft	18:59 0.57 ft	
Sat 23	00:48 3.51 ft	07:07 0.61 ft	13:00 3.76 ft	19:43 0.36 ft	
Sun 24	01:36 3.78 ft	07:55 0.36 ft	13:47 3.97 ft	20:26 0.15 ft	
Mon 25	02:17 4.08 ft	08:42 0.09 ft	14:28 4.17 ft	21:08 -0.05 ft	
Tue 26	02:55 4.38 ft	09:28 -0.15 ft	15:06 4.33 ft	21:50 -0.22 ft	
Wed 27	03:30 4.63 ft	10:14 -0.35 ft	15:44 4.42 ft	22:31 -0.34 ft	
Thu 28	04:07 4.82 ft	10:59 -0.49 ft	16:25 4.43 ft	23:13 -0.38 ft	
Fri 29	04:47 4.92 ft	11:44 -0.52 ft	17:10 4.37 ft	23:54 -0.34 ft	
Sat 30	05:32 4.92 ft	12:31 -0.45 ft	18:02 4.24 ft		
Sun 31		00:38 -0.21 ft	06:24 4.83 ft	13:21 -0.29 ft	19:00 4.10 ft

april 2013

Day	High	Low	High	Low	High
Mon 01		01:28 0.00 ft	07:23 4.69 ft	14:18 -0.07 ft	20:03 3.97 ft
Tue 02		02:29 0.24 ft	08:25 4.54 ft	15:24 0.11 ft	21:06 3.89 ft
Wed 03		03:41 0.41 ft	09:29 4.39 ft	16:32 0.17 ft	22:11 3.87 ft
Thu 04		04:54 0.42 ft	10:35 4.28 ft	17:36 0.10 ft	23:18 3.94 ft
Fri 05		05:59 0.30 ft	11:43 4.25 ft	18:33 -0.03 ft	
Sat 06	00:24 4.12 ft	06:58 0.12 ft	12:48 4.31 ft	19:25 -0.16 ft	
Sun 07	01:23 4.36 ft	07:52 -0.05 ft	13:45 4.40 ft	20:14 -0.25 ft	
Mon 08	02:14 4.59 ft	08:43 -0.19 ft	14:34 4.47 ft	21:01 -0.30 ft	
Tue 09	02:58 4.76 ft	09:32 -0.28 ft	15:19 4.48 ft	21:45 -0.29 ft	
Wed 10	03:39 4.83 ft	10:18 -0.31 ft	16:02 4.43 ft	22:28 -0.21 ft	
Thu 11	04:19 4.81 ft	11:02 -0.28 ft	16:44 4.30 ft	23:08 -0.07 ft	
Fri 12	04:57 4.70 ft	11:44 -0.17 ft	17:28 4.14 ft	23:46 0.12 ft	
Sat 13	05:36 4.53 ft	12:25 0.01 ft	18:13 3.95 ft		
Sun 14		00:23 0.35 ft	06:15 4.33 ft	13:05 0.23 ft	19:01 3.77 ft
Mon 15		00:58 0.60 ft	06:57 4.13 ft	13:47 0.48 ft	19:49 3.61 ft
Tue 16		01:34 0.83 ft	07:40 3.94 ft	14:35 0.70 ft	20:37 3.50 ft
Wed 17		02:17 1.04 ft	08:24 3.79 ft	15:32 0.86 ft	21:25 3.43 ft
Thu 18		03:23 1.18 ft	09:11 3.69 ft	16:32 0.90 ft	22:14 3.44 ft
Fri 19		04:37 1.17 ft	10:02 3.64 ft	17:26 0.82 ft	23:05 3.54 ft
Sat 20		05:39 1.02 ft	10:59 3.65 ft	18:15 0.67 ft	23:59 3.74 ft
Sun 21		06:33 0.78 ft	12:00 3.75 ft	19:01 0.47 ft	
Mon 22	00:49 4.03 ft	07:23 0.49 ft	12:57 3.91 ft	19:45 0.25 ft	
Tue 23	01:35 4.37 ft	08:12 0.18 ft	13:48 4.11 ft	20:30 0.05 ft	
Wed 24	02:17 4.72 ft	09:02 -0.11 ft	14:34 4.29 ft	21:15 -0.13 ft	
Thu 25	02:58 5.02 ft	09:51 -0.36 ft	15:19 4.42 ft	22:01 -0.27 ft	
Fri 26	03:40 5.22 ft	10:40 -0.54 ft	16:06 4.47 ft	22:48 -0.33 ft	
Sat 27	04:25 5.30 ft	11:28 -0.61 ft	16:57 4.44 ft	23:36 -0.31 ft	
Sun 28	05:15 5.26 ft	12:18 -0.58 ft	17:54 4.35 ft		
Mon 29		00:25 -0.19 ft	06:12 5.12 ft	13:09 -0.44 ft	18:56 4.26 ft
Tue 30		01:18 0.02 ft	07:14 4.92 ft	14:05 -0.24 ft	19:59 4.20 ft

may 2013

Day	High	Low	High	Low	High
Wed 01		02:19 0.26 ft	08:16 4.72 ft	15:06 -0.04 ft	21:00 4.17 ft
Thu 02		03:28 0.45 ft	09:17 4.52 ft	16:10 0.07 ft	22:01 4.18 ft
Fri 03		04:38 0.50 ft	10:18 4.34 ft	17:12 0.08 ft	23:02 4.23 ft
Sat 04		05:42 0.43 ft	11:21 4.21 ft	18:07 0.05 ft	
Sun 05	00:03 4.35 ft	06:40 0.31 ft	12:23 4.15 ft	18:58 0.01 ft	
Mon 06	00:59 4.50 ft	07:33 0.18 ft	13:20 4.15 ft	19:46 -0.00 ft	
Tue 07	01:50 4.66 ft	08:23 0.07 ft	14:11 4.18 ft	20:31 0.02 ft	
Wed 08	02:34 4.78 ft	09:11 -0.01 ft	14:56 4.19 ft	21:15 0.06 ft	
Thu 09	03:14 4.82 ft	09:56 -0.06 ft	15:39 4.17 ft	21:58 0.14 ft	
Fri 10	03:52 4.80 ft	10:40 -0.07 ft	16:21 4.11 ft	22:39 0.24 ft	
Sat 11	04:29 4.71 ft	11:21 -0.02 ft	17:04 4.01 ft	23:18 0.36 ft	
Sun 12	05:05 4.57 ft	12:01 0.08 ft	17:49 3.89 ft	23:55 0.51 ft	
Mon 13	05:42 4.40 ft	12:40 0.23 ft	18:35 3.76 ft		
Tue 14		00:31 0.68 ft	06:20 4.23 ft	13:19 0.41 ft	19:22 3.66 ft
Wed 15		01:05 0.85 ft	07:00 4.07 ft	14:00 0.58 ft	20:08 3.60 ft
Thu 16		01:43 1.01 ft	07:42 3.94 ft	14:46 0.72 ft	20:51 3.59 ft
Fri 17		02:34 1.14 ft	08:25 3.85 ft	15:39 0.79 ft	21:33 3.65 ft
Sat 18		03:48 1.18 ft	09:11 3.79 ft	16:35 0.77 ft	22:17 3.77 ft
Sun 19		04:58 1.07 ft	10:03 3.76 ft	17:28 0.65 ft	23:05 3.98 ft
Mon 20		05:58 0.84 ft	11:02 3.78 ft	18:17 0.47 ft	23:57 4.26 ft
Tue 21		06:52 0.54 ft	12:06 3.86 ft	19:05 0.27 ft	
Wed 22	00:50 4.59 ft	07:45 0.22 ft	13:08 4.01 ft	19:53 0.07 ft	
Thu 23	01:41 4.93 ft	08:37 -0.09 ft	14:05 4.19 ft	20:43 -0.10 ft	
Fri 24	02:30 5.23 ft	09:30 -0.36 ft	14:58 4.35 ft	21:35 -0.24 ft	
Sat 25	03:19 5.42 ft	10:22 -0.57 ft	15:50 4.44 ft	22:27 -0.33 ft	
Sun 26	04:09 5.49 ft	11:13 -0.69 ft	16:45 4.47 ft	23:19 -0.33 ft	
Mon 27	05:03 5.42 ft	12:03 -0.70 ft	17:44 4.46 ft		
Tue 28		00:12 -0.24 ft	06:01 5.26 ft	12:54 -0.60 ft	18:47 4.43 ft
Wed 29		01:06 -0.06 ft	07:02 5.04 ft	13:47 -0.42 ft	19:48 4.41 ft
Thu 30		02:05 0.18 ft	08:03 4.81 ft	14:44 -0.22 ft	20:46 4.41 ft
Fri 31		03:09 0.40 ft	09:00 4.55 ft	15:44 -0.04 ft	21:42 4.40 ft

june 2013

Day	High	Low	High	Low	High
Sat 01		04:16 0.52 ft	09:56 4.30 ft	16:43 0.08 ft	22:38 4.40 ft
Sun 02		05:20 0.53 ft	10:54 4.08 ft	17:38 0.14 ft	23:35 4.43 ft
Mon 03		06:18 0.47 ft	11:53 3.93 ft	18:28 0.19 ft	
Tue 04	00:31 4.49 ft	07:10 0.38 ft	12:51 3.87 ft	19:15 0.24 ft	
Wed 05	01:22 4.57 ft	08:00 0.29 ft	13:44 3.88 ft	20:01 0.28 ft	
Thu 06	02:08 4.64 ft	08:47 0.20 ft	14:32 3.92 ft	20:46 0.32 ft	
Fri 07	02:49 4.69 ft	09:33 0.13 ft	15:16 3.95 ft	21:29 0.36 ft	
Sat 08	03:27 4.69 ft	10:16 0.08 ft	15:59 3.95 ft	22:11 0.41 ft	
Sun 09	04:04 4.65 ft	10:58 0.06 ft	16:41 3.92 ft	22:52 0.46 ft	
Mon 10	04:39 4.56 ft	11:38 0.09 ft	17:24 3.86 ft	23:31 0.53 ft	
Tue 11	05:14 4.44 ft	12:15 0.17 ft	18:08 3.79 ft		
Wed 12		00:08 0.64 ft	05:49 4.30 ft	12:52 0.28 ft	18:52 3.74 ft
Thu 13		00:43 0.75 ft	06:25 4.17 ft	13:28 0.40 ft	19:34 3.72 ft
Fri 14		01:19 0.87 ft	07:03 4.06 ft	14:04 0.51 ft	20:13 3.76 ft
Sat 15		02:01 0.98 ft	07:46 3.97 ft	14:45 0.59 ft	20:51 3.86 ft
Sun 16		03:02 1.04 ft	08:32 3.89 ft	15:37 0.61 ft	21:32 4.01 ft
Mon 17		04:17 0.99 ft	09:23 3.83 ft	16:36 0.56 ft	22:19 4.21 ft
Tue 18		05:25 0.81 ft	10:20 3.80 ft	17:34 0.44 ft	23:13 4.44 ft
Wed 19		06:24 0.54 ft	11:26 3.81 ft	18:29 0.27 ft	
Thu 20	00:12 4.71 ft	07:20 0.23 ft	12:36 3.91 ft	19:24 0.08 ft	
Fri 21	01:12 5.00 ft	08:15 -0.08 ft	13:41 4.09 ft	20:19 -0.09 ft	
Sat 22	02:08 5.27 ft	09:09 -0.36 ft	14:40 4.29 ft	21:15 -0.25 ft	
Sun 23	03:02 5.45 ft	10:03 -0.59 ft	15:36 4.45 ft	22:10 -0.36 ft	
Mon 24	03:55 5.52 ft	10:55 -0.74 ft	16:32 4.55 ft	23:04 -0.40 ft	
Tue 25	04:50 5.46 ft	11:45 -0.79 ft	17:30 4.59 ft	23:57 -0.33 ft	
Wed 26	05:47 5.29 ft	12:35 -0.72 ft	18:30 4.60 ft		
Thu 27		00:50 -0.16 ft	06:45 5.06 ft	13:25 -0.55 ft	19:28 4.59 ft
Fri 28		01:46 0.09 ft	07:42 4.78 ft	14:17 -0.30 ft	20:24 4.56 ft
Sat 29		02:46 0.35 ft	08:37 4.49 ft	15:12 -0.05 ft	21:17 4.51 ft
Sun 30		03:50 0.55 ft	09:30 4.20 ft	16:09 0.17 ft	22:09 4.43 ft

july 2013

Day	High	Low	High	Low	High
Mon 01		04:53 0.63 ft	10:25 3.93 ft	17:04 0.33 ft	23:02 4.37 ft
Tue 02		05:52 0.63 ft	11:22 3.74 ft	17:56 0.43 ft	23:57 4.35 ft
Wed 03		06:45 0.57 ft	12:20 3.65 ft	18:45 0.49 ft	
Thu 04	00:50 4.38 ft	07:34 0.48 ft	13:16 3.66 ft	19:31 0.51 ft	
Fri 05	01:39 4.44 ft	08:21 0.38 ft	14:07 3.74 ft	20:17 0.51 ft	
Sat 06	02:24 4.52 ft	09:07 0.28 ft	14:53 3.83 ft	21:02 0.49 ft	
Sun 07	03:04 4.57 ft	09:50 0.19 ft	15:36 3.90 ft	21:45 0.47 ft	
Mon 08	03:41 4.59 ft	10:32 0.12 ft	16:17 3.93 ft	22:28 0.45 ft	
Tue 09	04:16 4.56 ft	11:11 0.09 ft	16:57 3.93 ft	23:08 0.46 ft	
Wed 10	04:49 4.48 ft	11:49 0.10 ft	17:37 3.92 ft	23:46 0.51 ft	
Thu 11	05:20 4.38 ft	12:24 0.16 ft	18:15 3.91 ft		
Fri 12		00:22 0.58 ft	05:53 4.27 ft	12:56 0.24 ft	18:53 3.93 ft
Sat 13		00:58 0.67 ft	06:30 4.16 ft	13:28 0.33 ft	19:30 4.00 ft
Sun 14		01:38 0.77 ft	07:13 4.05 ft	14:01 0.41 ft	20:10 4.11 ft
Mon 15		02:31 0.85 ft	08:02 3.96 ft	14:44 0.48 ft	20:54 4.25 ft
Tue 16		03:43 0.87 ft	08:55 3.87 ft	15:49 0.52 ft	21:44 4.40 ft
Wed 17		04:56 0.76 ft	09:54 3.80 ft	17:00 0.46 ft	22:41 4.56 ft
Thu 18		06:00 0.53 ft	11:01 3.79 ft	18:04 0.32 ft	23:45 4.74 ft
Fri 19		06:59 0.24 ft	12:15 3.88 ft	19:04 0.14 ft	
Sat 20	00:52 4.97 ft	07:55 -0.05 ft	13:25 4.08 ft	20:02 -0.06 ft	
Sun 21	01:54 5.21 ft	08:50 -0.33 ft	14:27 4.33 ft	20:59 -0.24 ft	
Mon 22	02:49 5.40 ft	09:44 -0.57 ft	15:23 4.56 ft	21:55 -0.37 ft	
Tue 23	03:42 5.47 ft	10:35 -0.73 ft	16:16 4.71 ft	22:48 -0.43 ft	
Wed 24	04:35 5.41 ft	11:24 -0.78 ft	17:11 4.79 ft	23:40 -0.37 ft	
Thu 25	05:28 5.24 ft	12:11 -0.71 ft	18:06 4.79 ft		
Fri 26		00:31 -0.21 ft	06:22 4.98 ft	12:58 -0.51 ft	19:01 4.73 ft
Sat 27		01:24 0.05 ft	07:17 4.68 ft	13:46 -0.23 ft	19:54 4.64 ft
Sun 28		02:19 0.35 ft	08:10 4.37 ft	14:36 0.09 ft	20:45 4.52 ft
Mon 29		03:19 0.61 ft	09:02 4.07 ft	15:30 0.39 ft	21:36 4.38 ft
Tue 30		04:21 0.77 ft	09:55 3.82 ft	16:27 0.62 ft	22:26 4.26 ft
Wed 31		05:22 0.81 ft	10:50 3.63 ft	17:22 0.74 ft	23:20 4.18 ft

august 2013

Day	High	Low	High	Low	High
Thu 01		06:16 0.76 ft	11:49 3.55 ft	18:14 0.77 ft	
Fri 02	00:16 4.18 ft	07:06 0.67 ft	12:47 3.58 ft	19:02 0.74 ft	
Sat 03	01:09 4.25 ft	07:53 0.55 ft	13:40 3.70 ft	19:49 0.67 ft	
Sun 04	01:57 4.37 ft	08:37 0.41 ft	14:27 3.85 ft	20:35 0.58 ft	
Mon 05	02:39 4.48 ft	09:20 0.29 ft	15:09 3.98 ft	21:19 0.48 ft	
Tue 06	03:16 4.55 ft	10:02 0.17 ft	15:49 4.09 ft	22:02 0.41 ft	
Wed 07	03:50 4.58 ft	10:40 0.10 ft	16:25 4.15 ft	22:43 0.36 ft	
Thu 08	04:21 4.55 ft	11:17 0.06 ft	17:00 4.20 ft	23:23 0.35 ft	
Fri 09	04:52 4.48 ft	11:52 0.08 ft	17:34 4.23 ft		
Sat 10		00:01 0.38 ft	05:24 4.38 ft	12:24 0.13 ft	18:09 4.28 ft
Sun 11		00:39 0.45 ft	06:02 4.25 ft	12:56 0.22 ft	18:48 4.33 ft
Mon 12		01:20 0.55 ft	06:48 4.12 ft	13:30 0.33 ft	19:33 4.40 ft
Tue 13		02:11 0.67 ft	07:41 3.99 ft	14:13 0.45 ft	20:24 4.47 ft
Wed 14		03:20 0.75 ft	08:39 3.88 ft	15:19 0.56 ft	21:19 4.53 ft
Thu 15		04:35 0.71 ft	09:41 3.81 ft	16:40 0.56 ft	22:21 4.61 ft
Fri 16		05:42 0.52 ft	10:51 3.82 ft	17:49 0.43 ft	23:29 4.72 ft
Sat 17		06:41 0.26 ft	12:06 3.94 ft	18:51 0.22 ft	
Sun 18	00:39 4.90 ft	07:37 -0.03 ft	13:15 4.20 ft	19:49 0.00 ft	
Mon 19	01:42 5.11 ft	08:31 -0.29 ft	14:15 4.50 ft	20:45 -0.20 ft	
Tue 20	02:37 5.28 ft	09:23 -0.50 ft	15:08 4.76 ft	21:39 -0.33 ft	
Wed 21	03:28 5.35 ft	10:12 -0.63 ft	15:58 4.93 ft	22:31 -0.39 ft	
Thu 22	04:16 5.28 ft	10:59 -0.65 ft	16:47 4.99 ft	23:22 -0.34 ft	
Fri 23	05:06 5.10 ft	11:44 -0.55 ft	17:37 4.95 ft		
Sat 24		00:10 -0.17 ft	05:56 4.83 ft	12:28 -0.33 ft	18:28 4.83 ft
Sun 25		00:59 0.08 ft	06:49 4.53 ft	13:12 -0.02 ft	19:19 4.67 ft
Mon 26		01:49 0.38 ft	07:41 4.23 ft	13:58 0.33 ft	20:09 4.48 ft
Tue 27		02:44 0.68 ft	08:33 3.96 ft	14:48 0.67 ft	20:58 4.30 ft
Wed 28		03:45 0.89 ft	09:25 3.74 ft	15:46 0.92 ft	21:48 4.14 ft
Thu 29		04:46 0.97 ft	10:19 3.59 ft	16:46 1.04 ft	22:41 4.04 ft
Fri 30		05:43 0.94 ft	11:17 3.53 ft	17:42 1.04 ft	23:38 4.02 ft
Sat 31		06:34 0.83 ft	12:16 3.59 ft	18:33 0.94 ft	

september 2013

Day	High	Low	High	Low	High
Sun 01	00:34 4.10 ft	07:20 0.68 ft	13:10 3.74 ft	19:21 0.80 ft	
Mon 02	01:25 4.23 ft	08:04 0.52 ft	13:58 3.94 ft	20:07 0.64 ft	
Tue 03	02:09 4.38 ft.	08:46 0.36 ft	14:39 4.14 ft	20:52 0.48 ft	
Wed 04	02:46 4.50 ft	09:27 0.22 ft	15:16 4.32 ft	21:35 0.34 ft	
Thu 05	03:20 4.57 ft	10:06 0.11 ft	15:50 4.46 ft	22:18 0.23 ft	
Fri 06	03:52 4.58 ft	10:43 0.04 ft	16:22 4.57 ft	23:00 0.17 ft	
Sat 07	04:24 4.54 ft	11:19 0.02 ft	16:54 4.64 ft	23:40 0.16 ft	
Sun 08	04:59 4.45 ft	11:54 0.06 ft	17:31 4.68 ft		
Mon 09		00:22 0.21 ft	05:41 4.31 ft	12:30 0.15 ft	18:14 4.69 ft
Tue 10		01:06 0.33 ft	06:31 4.16 ft	13:09 0.29 ft	19:06 4.67 ft
Wed 11		01:59 0.48 ft	07:29 4.01 ft	13:58 0.47 ft	20:03 4.64 ft
Thu 12		03:05 0.60 ft	08:33 3.91 ft	15:08 0.63 ft	21:05 4.61 ft
Fri 13		04:18 0.60 ft	09:39 3.87 ft	16:29 0.65 ft	22:10 4.60 ft
Sat 14		05:25 0.46 ft	10:48 3.92 ft	17:38 0.51 ft	23:19 4.65 ft
Sun 15		06:24 0.22 ft	12:00 4.09 ft	18:40 0.29 ft	
Mon 16	00:28 4.77 ft	07:19 -0.03 ft	13:05 4.38 ft	19:36 0.06 ft	
Tue 17	01:29 4.94 ft	08:10 -0.25 ft	14:01 4.68 ft	20:31 -0.13 ft	
Wed 18	02:22 5.06 ft	08:59 -0.41 ft	14:51 4.94 ft	21:23 -0.26 ft	
Thu 19	03:10 5.10 ft	09:47 -0.48 ft	15:38 5.09 ft	22:13 -0.31 ft	
Fri 20	03:56 5.02 ft	10:32 -0.46 ft	16:22 5.11 ft	23:02 -0.26 ft	
Sat 21	04:42 4.85 ft	11:16 -0.32 ft	17:07 5.02 ft	23:48 -0.12 ft	
Sun 22	05:30 4.60 ft	11:58 -0.10 ft	17:53 4.85 ft		
Mon 23		00:33 0.11 ft	06:19 4.33 ft	12:39 0.20 ft	18:40 4.63 ft
Tue 24		01:19 0.38 ft	07:11 4.06 ft	13:20 0.53 ft	19:29 4.40 ft
Wed 25		02:09 0.67 ft	08:04 3.83 ft	14:05 0.85 ft	20:18 4.19 ft
Thu 26		03:05 0.91 ft	08:56 3.66 ft	14:59 1.10 ft	21:08 4.03 ft
Fri 27		04:07 1.03 ft	09:49 3.55 ft	16:04 1.24 ft	21:59 3.92 ft
Sat 28		05:05 1.02 ft	10:43 3.53 ft	17:06 1.22 ft	22:54 3.88 ft
Sun 29		05:58 0.91 ft	11:40 3.60 ft	18:01 1.08 ft	23:51 3.93 ft
Mon 30		06:44 0.75 ft	12:34 3.77 ft	18:51 0.89 ft	

october 2013

Day	High	Low	High	Low	High
Tue 01	00:44 4.04 ft	07:27 0.56 ft	13:22 4.00 ft	19:37 0.67 ft	
Wed 02	01:31 4.19 ft	08:08 0.37 ft	14:03 4.26 ft	20:23 0.45 ft	
Thu 03	02:11 4.34 ft	08:49 0.20 ft	14:40 4.51 ft	21:08 0.24 ft	
Fri 04	02:47 4.45 ft	09:29 0.06 ft	15:14 4.73 ft	21:52 0.07 ft	
Sat 05	03:22 4.51 ft	10:09 -0.03 ft	15:47 4.90 ft	22:37 -0.05 ft	
Sun 06	03:59 4.50 ft	10:48 -0.08 ft	16:23 5.00 ft	23:21 -0.11 ft	
Mon 07	04:39 4.43 ft	11:28 -0.06 ft	17:03 5.01 ft		
Tue 08		00:06 -0.08 ft	05:25 4.30 ft	12:10 0.03 ft	17:51 4.95 ft
Wed 09		00:54 0.03 ft	06:21 4.15 ft	12:56 0.19 ft	18:48 4.84 ft
Thu 10		01:48 0.19 ft	07:26 4.02 ft	13:51 0.40 ft	19:51 4.72 ft
Fri 11		02:51 0.34 ft	08:32 3.95 ft	15:01 0.57 ft	20:56 4.61 ft
Sat 12		04:00 0.38 ft	09:37 3.96 ft	16:18 0.61 ft	22:01 4.54 ft
Sun 13		05:06 0.28 ft	10:44 4.05 ft	17:26 0.49 ft	23:07 4.51 ft
Mon 14		06:04 0.10 ft	11:50 4.23 ft	18:27 0.29 ft	
Tue 15	00:12 4.55 ft	06:57 -0.09 ft	12:51 4.48 ft	19:23 0.09 ft	
Wed 16	01:12 4.62 ft	07:47 -0.24 ft	13:45 4.75 ft	20:16 -0.08 ft	
Thu 17	02:04 4.69 ft	08:35 -0.33 ft	14:33 4.95 ft	21:06 -0.19 ft	
Fri 18	02:52 4.70 ft	09:20 -0.34 ft	15:16 5.06 ft	21:55 -0.24 ft	
Sat 19	03:36 4.63 ft	10:05 -0.27 ft	15:57 5.05 ft	22:41 -0.21 ft	
Sun 20	04:20 4.50 ft	10:47 -0.14 ft	16:37 4.94 ft	23:26 -0.11 ft	
Mon 21	05:04 4.30 ft	11:28 0.06 ft	17:18 4.75 ft		
Tue 22		00:08 0.06 ft	05:52 4.08 ft	12:07 0.30 ft	18:01 4.52 ft
Wed 23		00:51 0.28 ft	06:42 3.86 ft	12:46 0.57 ft	18:47 4.29 ft
Thu 24		01:36 0.53 ft	07:34 3.68 ft	13:26 0.83 ft	19:36 4.08 ft
Fri 25		02:25 0.75 ft	08:25 3.55 ft	14:12 1.06 ft	20:25 3.91 ft
Sat 26		03:21 0.90 ft	09:16 3.48 ft	15:14 1.21 ft	21:13 3.79 ft
Sun 27		04:20 0.94 ft	10:06 3.47 ft	16:23 1.23 ft	22:03 3.73 ft
Mon 28		05:14 0.86 ft	10:58 3.55 ft	17:24 1.11 ft	22:56 3.72 ft
Tue 29		06:03 0.71 ft	11:49 3.72 ft	18:17 0.89 ft	23:51 3.78 ft
Wed 30		06:46 0.51 ft	12:38 3.97 ft	19:06 0.63 ft	
Thu 31	00:43 3.90 ft	07:28 0.31 ft	13:21 4.26 ft	19:53 0.36 ft	

november 2013

Day	High	Low	High	Low	High
Fri 01	01:30 4.05 ft	08:10 0.12 ft	14:01 4.57 ft	20:40 0.10 ft	
Sat 02	02:13 4.20 ft	08:53 -0.05 ft	14:39 4.86 ft	21:28 -0.14 ft	
Sun 03	01:55 4.31 ft	08:36 -0.19 ft	14:18 5.07 ft	21:15 -0.32 ft	
Mon 04	02:37 4.36 ft	09:22 -0.27 ft	14:59 5.18 ft	22:03 -0.42 ft	
Tue 05	03:23 4.33 ft	10:08 -0.28 ft	15:45 5.18 ft	22:51 -0.43 ft	
Wed 06	04:15 4.25 ft	10:55 -0.21 ft	16:37 5.07 ft	23:41 -0.36 ft	
Thu 07	05:15 4.13 ft	11:46 -0.06 ft	17:37 4.90 ft		
Fri 08		00:34 -0.21 ft	06:21 4.05 ft	12:43 0.15 ft	18:42 4.72 ft
Sat 09		01:34 -0.06 ft	07:26 4.02 ft	13:50 0.35 ft	19:45 4.55 ft
Sun 10		02:38 0.04 ft	08:28 4.04 ft	15:02 0.44 ft	20:47 4.39 ft
Mon 11		03:42 0.02 ft	09:30 4.12 ft	16:10 0.38 ft	21:49 4.26 ft
Tue 12		04:40 -0.06 ft	10:32 4.25 ft	17:11 0.24 ft	22:52 4.19 ft
Wed 13		05:33 -0.15 ft	11:31 4.42 ft	18:07 0.08 ft	23:51 4.18 ft
Thu 14		06:22 -0.22 ft	12:25 4.60 ft	18:58 -0.06 ft	
Fri 15	00:44 4.20 ft	07:09 -0.24 ft	13:12 4.74 ft	19:48 -0.15 ft	
Sat 16	01:32 4.20 ft	07:54 -0.22 ft	13:55 4.81 ft	20:35 -0.21 ft	
Sun 17	02:17 4.18 ft	08:38 -0.15 ft	14:35 4.80 ft	21:20 -0.22 ft	
Mon 18	02:59 4.10 ft	09:21 -0.05 ft	15:13 4.71 ft	22:03 -0.17 ft	
Tue 19	03:42 3.98 ft	10:01 0.08 ft	15:51 4.56 ft	22:45 -0.07 ft	
Wed 20	04:27 3.83 ft	10:40 0.24 ft	16:30 4.37 ft	23:25 0.08 ft	
Thu 21	05:14 3.67 ft	11:18 0.42 ft	17:11 4.17 ft		
Fri 22		00:06 0.26 ft	06:03 3.54 ft	11:55 0.62 ft	17:55 3.98 ft
Sat 23		00:47 0.45 ft	06:52 3.44 ft	12:33 0.81 ft	18:39 3.81 ft
Sun 24		01:34 0.60 ft	07:38 3.39 ft	13:21 0.97 ft	19:23 3.68 ft
Mon 25		02:26 0.69 ft	08:23 3.40 ft	14:29 1.05 ft	20:08 3.58 ft
Tue 26		03:22 0.68 ft	09:08 3.48 ft	15:40 1.00 ft	20:55 3.53 ft
Wed 27		04:14 0.58 ft	09:55 3.64 ft	16:40 0.81 ft	21:49 3.52 ft
Thu 28		05:01 0.42 ft	10:44 3.87 ft	17:33 0.54 ft	22:48 3.58 ft
Fri 29		05:47 0.21 ft	11:34 4.17 ft	18:24 0.24 ft	23:47 3.71 ft
Sat 30		06:33 -0.00 ft	12:23 4.50 ft	19:14 -0.07 ft	

december 2013

Day	High	Low	High	Low	High
Sun 01	00:41 3.88 ft	07:21 -0.21 ft	13:09 4.81 ft	20:05 -0.35 ft	
Mon 02	01:31 4.05 ft	08:10 -0.39 ft	13:55 5.06 ft	20:55 -0.58 ft	
Tue 03	02:20 4.18 ft	09:00 -0.52 ft	14:42 5.19 ft	21:46 -0.74 ft	
Wed 04	03:11 4.23 ft	09:51 -0.58 ft	15:32 5.19 ft	22:35 -0.81 ft	
Thu 05	04:06 4.22 ft	10:42 -0.55 ft	16:27 5.08 ft	23:26 -0.77 ft	
Fri 06	05:06 4.17 ft	11:35 -0.42 ft	17:27 4.89 ft		
Sat 07		00:17 -0.65 ft	06:10 4.13 ft	12:31 -0.21 ft	18:29 4.66 ft
Sun 08		01:12 -0.47 ft	07:11 4.11 ft	13:33 0.03 ft	19:29 4.42 ft
Mon 09		02:12 -0.30 ft	08:10 4.10 ft	14:42 0.20 ft	20:27 4.18 ft
Tue 10		03:13 -0.19 ft	09:08 4.11 ft	15:49 0.25 ft	21:26 3.96 ft
Wed 11		04:12 -0.14 ft	10:07 4.14 ft	16:51 0.19 ft	22:26 3.79 ft
Thu 12		05:06 -0.13 ft	11:06 4.20 ft	17:47 0.09 ft	23:26 3.71 ft
Fri 13		05:56 -0.13 ft	12:01 4.29 ft	18:39 -0.01 ft	
Sat 14	00:22 3.71 ft	06:43 -0.12 ft	12:50 4.38 ft	19:28 -0.11 ft	
Sun 15	01:12 3.74 ft	07:29 -0.10 ft	13:34 4.44 ft	20:14 -0.18 ft	
Mon 16	01:58 3.78 ft	08:13 -0.08 ft	14:14 4.46 ft	20:58 -0.23 ft	
Tue 17	02:40 3.79 ft	08:56 -0.05 ft	14:52 4.43 ft	21:41 -0.25 ft	
Wed 18	03:22 3.75 ft	09:38 -0.00 ft	15:29 4.35 ft	22:21 -0.22 ft	
Thu 19	04:04 3.68 ft	10:17 0.07 ft	16:05 4.23 ft	23:00 -0.15 ft	
Fri 20	04:47 3.58 ft	10:54 0.18 ft	16:41 4.08 ft	23:37 -0.03 ft	
Sat 21	05:30 3.49 ft	11:30 0.31 ft	17:18 3.91 ft		
Sun 22		00:13 0.11 ft	06:13 3.42 ft	12:04 0.47 ft	17:55 3.76 ft
Mon 23		00:48 0.25 ft	06:54 3.39 ft	12:41 0.61 ft	18:33 3.62 ft
Tue 24		01:25 0.37 ft	07:33 3.41 ft	13:31 0.73 ft	19:16 3.50 ft
Wed 25		02:12 0.44 ft	08:13 3.49 ft	14:45 0.78 ft	20:03 3.42 ft
Thu 26		03:12 0.43 ft	08:57 3.63 ft	15:58 0.67 ft	20:57 3.36 ft
Fri 27		04:13 0.33 ft	09:49 3.82 ft	17:00 0.44 ft	21:59 3.36 ft
Sat 28		05:08 0.15 ft	10:47 4.06 ft	17:56 0.15 ft	23:08 3.45 ft
Sun 29		06:02 -0.08 ft	11:48 4.36 ft	18:50 -0.17 ft	
Mon 30	00:14 3.64 ft	06:55 -0.32 ft	12:45 4.68 ft	19:44 -0.48 ft	
Tue 31	01:13 3.87 ft	07:50 -0.54 ft	13:38 4.94 ft	20:36 -0.76 ft	

Archival Texts by Ben Franklin

EXCERPTS FROM ARCHIVAL POOR RICHARD ALMANACKS

In 1732, I first published my Almanack under the name of "Richard Sanders"; it was continued by me about twenty-five years, and commonly called "Poor Richard's Almanack". I endeavored to make both entertaining and useful, and it accordingly came to be in such demand that I reaped considerable profit from it; vending annually near ten thousand. And observing that it was generally read, (scarce any neighborhood in the province being without it,) I considered it as a proper vehicle for conveying instruction among the common people, who bought scarcely any other books. I therefore filled all the little spaces that occurred between the remarkable days in the Calendar, with proverbial sentences, chiefly such as inculcated industry and frugality, as the means of procuring wealth, and thereby securing virtue; it being more difficult for a man in want to act always honestly, as (to use here one of those proverbs) "it is hard for an empty sack to stand upright." These proverbs, which contained the wisdom of many ages and nations, I assembled and formed into a connected discourse prefixed of the Almanack 1757, as the harangue of a wise old man to the people attending an auction: the bringing all these scattered counsels thus into a focus, enabled them to make greater impressions. The piece being universally approved, was copied in all the newspapers of the American Continent, reprinted in Britain on a large sheet of paper to be stuck up in houses; two translations were made of it in France, and great numbers bought by the clergy and gentry, to distribute gratis among their poor parishioners and tenants. In Pennsylvania, as it discouraged useless expense in foreign superfluities, some thought it had its share of influence in producing the growing plenty of money which was observable for several years after its publication.

"SUDDEN POWER IS APT TO BE INSOLENT, SUDDEN LIBERTY SAUCY; THAT BEHAVES BEST WHICH HAS GROWN GRADUALLY."

Be rarely warm in Censure or in Praise;
Few Men deserve our Passion either ways:
For half the World but floats 'twixt Good and Ill,
As Chance disposes Objects, these the Will;
'Tis but a see-saw Game, where Virtue now
Mounts about Vice, and then sinks down as low.
Besides, the Wise still hold it for a Rule,
To Trust the Judgement most, that seems most cool.

"I have constantly interpers'd in every Vacancy, Moral Hints, Wise Sayings, and Maxims of Thrift, tending to impress the Benefits arising from Honesty, Sobriety, Industry and Frugality; which is those has duly observed, it is highly probably thou are wiser and richer many fold more that the Pence my Labours have cost thee. Howbeit, I shall not therefore raise my Price because thou are better able to pay; but being thankful for past Favours, shall endeavor to make my little Book more worthy thy Regard, by adding those Recipes which were intended to Cure the Mind, some valuable Ones regarding the Health of the Body. They are recommended by the skillful, and by successful Practice. I wish a Blessing may attend the Use of them, and to thee all Happiness, being."

MONTH-SPECIFIC QUOTATIONS

OCTOBER

Singularity in the right, hath ruined many: Happy those who are convinced of the general Opinion.

To serve the Publick faithfully, and at the same time please it entirely is impracticable.

JUNE

He that's content hath enough.
He that complains has too much.

DECEMBER

With bounteous cheer
Conclude the year.

MARCH

To HIM intrust thy Slumbers, and prepare
The fragrant Incense of they Ev'ning Prayer.
But first tread back the Day, with Search severe,
And Conscience, chiding or applauding, hear.
Review each Step; Where, acting, did I err?
Omitting, where? Guilt either Way inter.
Labour this Point, and while they Frailties last,
Still let each following Day correct last.

Tis a well spent penny that saves a groat.

Many Foxes grow grey, but few grow good.

Presumption first blinds a Man,
then sets him a running.

Introduction to the Almanac Zodiac Calendar

by CHRISTOPHER CHEMSAK

As far as all-around farm tools go, some would maintain that a calendar is among the most indispensable—right up there with a digging spade and a pickup truck. As a lens into the future, a guide for the present, and a record of the past, a good planting calendar can greatly assist in attempts to organize, document, schedule, sow, harvest, slaughter, prune, party, settle up, or settle in. Still, the calendar cannot predict the inevitable broken-down tractor, sick calf, or infested broccoli plants. One may find, however, that efficiently planning seasonal and daily operations with calendar-inspired forethought can allow for greater flexibility and less stress when unanticipated situations do arise. Add meticulous record-keeping to the mix and the practicality of this singular tool actually becomes quite time consuming.

For many cultures, the use of calendar-like devices to guide agricultural endeavors goes back to time immemorial. In the United States, since the time of Ben Franklin's Poor Richard's Almanack, annual farmer's almanacs have consolidated forecasts on weather, moon phases, implications of the zodiac, and sunrise and sunset times to assist farmers in tending their vegetable gardens. For many generations before internet and television, when most families produced their own food supply, the almanac's predictions were held in high regard and used to determine monthly planting schedules, which were often based on the favorability or unfavorability of the Moon's position in the zodiac and/or its stage in its cycle.

Today, many farmers and gardeners still heed the same factors and calculations for planting by the Moon. The "rules" and understandings have been passed down mostly through folkloric tradition. All of these guidelines require a bit of background knowledge as well as an open mind, so an attempt at a shorthand version seems unfitting. An additional, excellent resource is the one-of-a-kind book Raising with the Moon by Jack R. Pyle and Taylor Reese.

Similar practices and explanations for planting by the Moon have come out of the Biodynamic movement, deriving from the teachings of 20th Century Austrian philosopher/mystic, Rudolf Steiner. Steiner's 1924 series of lectures on agriculture contain all the instructions and principles that have standardized the methods of Biodynamic farmers and gardeners across the world. Though the Biodynamic planting calendar, known as the Stella Natura Calendar, was not a direct product of Steiner's work and is not considered necessary to the practice of Biodynamics, its potential functionality is unmistakable for anyone interested in following lunar rhythms in farming.

The adapted Stella Natura Calendar contains the Moon's phase, and its relative position in the zodiac.....

THE CIRCLE OF THE MONTHS
(From the Kalender of Shepherdes, 1503)

PREFACE.

I've cum forward wunst more to make obedience to the publick, and to circumbobbolato my idees to obleege my friend, the Kurnel, and to mortalize him as long as I live. Sum unbeleaving lubbers has got it that my friend is not in the minds, and I have got him to send word that he is alive, and to send me his Orto Graff, which is here disjoined for to shoo that he is above water yit:

"BEN HARDING ESQ.—DEAR CUR, I am alive. Take keer of my dog Growler and my wife.

Yourn, till deth,

This goes to shoo that my frend is alive; bekase why? If wa'nt alive he couldn't rite his name. As my frend the Kurnel has put down his Orto Graff for the public to reed, I've got the pictur folks to make mine too.

So here is both our names put down, and we haint done this to enny other almynick; but this is done to shoo that it is the jenuwine one, and 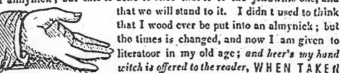 that we will stand to it. I didn't used to think that I wood ever be put into an almynick; but the times is changed, and now I am given to literatoor in my old age; *and heer's my hand witch is offered to the reader,* WHEN TAKEN TO BE WELL SHAKEN, I've got a bit of a yarn to tell the public about myself. I'm goin to git marrid. The gal of my choise is Susannah Twig, who lives on the left side of Yell Hollow, jist above the fork of Roaring River. I'm going to do this so as to raze up a hare to take keer of the almynick when I am ded and gone, and it is all for the public good, and to obleege my frend the Kurnel.

May, 1842.

TIDES AND THE LUNAR CYCLE *by* CHRISTEN RIPLEY

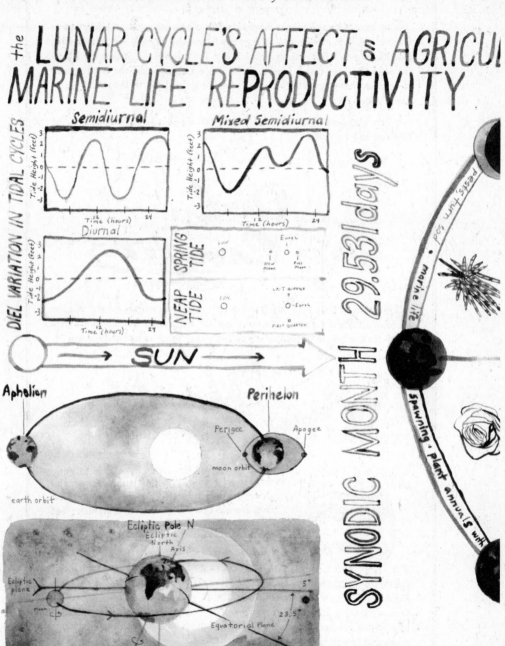

the LUNAR CYCLE'S AFFECT on AGRICUL
MARINE LIFE REPRODUCTIVITY

DIEL VARIATION IN TIDAL CYCLES

Semidiurnal

Mixed Semidiurnal

Diurnal

SPRING TIDE

NEAP TIDE

SUN

SYNODIC MONTH 29.531 days

Aphelion

Perihelon

Perigee

Apogee

moon orbit

earth orbit

Ecliptic Pole N
Ecliptic North Axis

Ecliptic plane

5°

23.5°

Equatorial Plane

moon

Ecliptic South

marine life

spawning · plant annuals with

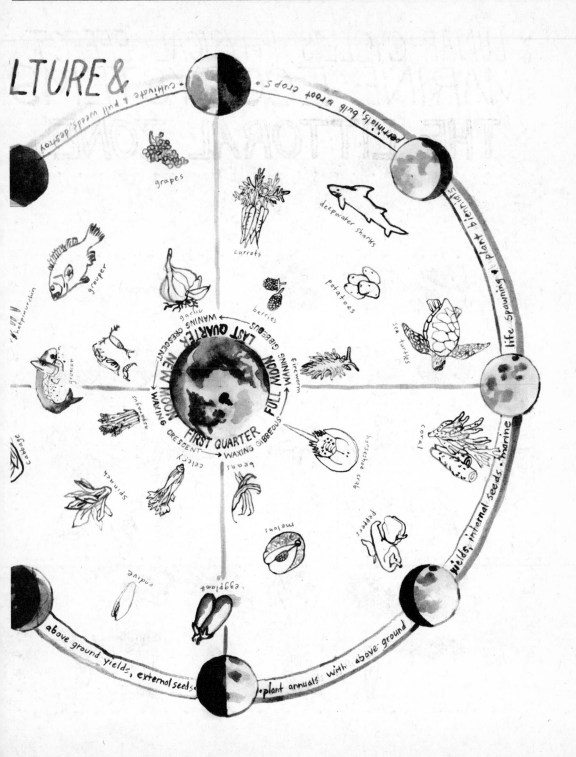

the LUNAR CYCLE'S PHYSICAL EFFECT on MARINE ECOSYSTEMS
THE LITTORAL ZONE

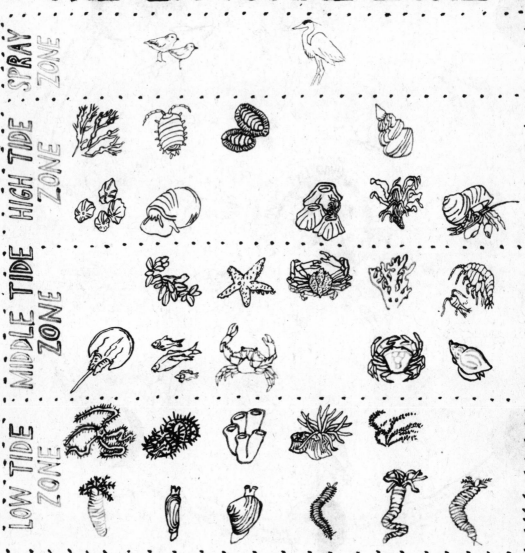

WATERSHED &
HYDROLOGIC & CARBON CYCLE

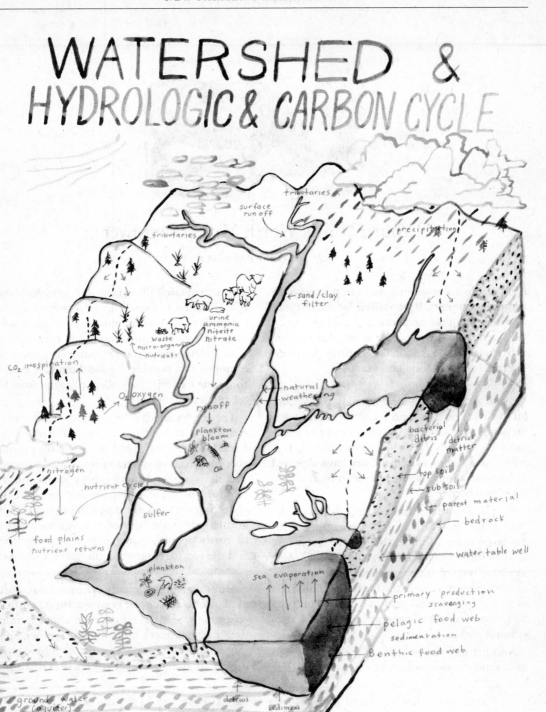

Your Farm Is Worth More Than Ever

by ROBERT RODALE

Agriculture is known to have been practiced for about 10,000 years. If we assume for the purpose of discussion that agriculture is exactly 10,000 years old, then it is accurate to say that farmers and their farms functioned entirely on their internal resources for 9,900 years.

By internal resources, I mean the land itself, the sun, air, rainfall, plants, animals, people and all the other physical and human resources that are within the immediate environment of every farm. Historically, farms have been uniquely self-reliant production systems. They have supplied almost everything they have needed from within their own borders.

Think about the land itself, for example. Within the silt portion of all agricultural land are large stores of minerals that are frequently unavailable to plants. Those minerals become available little by little, and also circulate within the cropping systems of the farm. For thousands of years, that internal resource of the land was relied on to supply all the mineral needs of agricultural plants.

Another internal resource is the air. It is rich in carbon dioxide, which is needed by plants. The air contains 78 percent nitrogen. Although that nitrogen is locked up in a fashion similar to the large supply of minerals in farm soils, some plants and microorganisms have the key to unlock that important resource. Legumes, for example, can provide a good home for certain root-zone bacteria that collect nitrogen from the air and lets plant use it.

The energy needs of agriculture are also interesting to examine from a historical perspective. Until fairly recently, agriculture was the primary energy-collecting system of developed societies. The sun was the primary energy source, but agricultural plants and animals were the agents for collecting the sun's rays and embodying them in a form that could be used throughout the year to satisfy people's own energy needs. The farm and the sun together were, therefore, a uniquely economical energy source.

Moisture from rain is also an internal resource of a farm. So are the plants grown, the domesticated animals, and the information about farming itself in the farmer's head. Taken together, all of those important productive elements have—over the past 10,000 years—provided a firm base upon

ORIGINALLY PUBLISHED IN THE JANUARY 1986 ISSUE OF THE NEW FARM MAGAZINE.
REPRINTED WITH PERMISSION FROM RODALE INSTITUTE. WWW.RODALEINSTITUTE.ORG

which people were able to build a civilized way of life.

But about 100 years ago, agriculture began to move beyond its vast and useful internal resources into a production system based also on external inputs like fertilizer, pesticides and fossil fuels. What caused this change? Was it the pressure of increasing population? Or did advancing science open doors to an awareness of agricultural processes that, in turn, led to the development of external input production?

I think it was both. Even 200 years ago, there was concern that population was expanding beyond the potential of agriculture as it was then practiced to feed everyone. Jethro Tull, the inventor of row cropping and thorough tillage and weeding, warned in the early 1700s that unless farmers adopted his advanced methods, people would either starve or start eating each other.

Milking machine

And about 100 years ago, the great German chemist Justus von Leibig formulated his law of the minimum, which states that crop yield is limited by the level of the nutrients present in the smallest amount. That insight, and other contributions of von Leibig, opened the way for the creation of the modern fertilizer industry.

Who else contributed to the idea that the technique of increasing crop production by bringing in materials from outside the farm? That would be a long list indeed. On it would be many chemists, plant breeders, soil experts, engineers and developers of oil and gas resources.

One hundred years is a short interval in the 10,000-year history of agriculture. Yet in that time, the production of food for all the peoples of the world has switched dramatically from total reliance on the internal resources of farms to systems of external inputs. While that change has enabled farms in the United States and other developed countries to produce ever larger amounts of food, I contend that it has distracted our attention from the enormous value, and indeed, the primacy of the internal resources of agriculture.

And the systems using external inputs have done farms more than merely cloud our vision of the value of internal resources. The external inputs, themselves, often reduce the usefulness of a farm's internal resources.

Nitrogen is one of the best examples. When external sources of nitrogen are put into the soil, the rich population of microbes that capture nitrogen from the air does its work less effectively. Often, it ceases to function. And likewise, soil minerals become available more slowly and are recycled much less effectively when input-intensive farming systems are used. And irrigation systems can make rainfall and other environmental constraints irrelevant to crop production, at least in the short run.

Those of us now advocating regenerative farming systems contend that the pendulum should start swinging the other way. We are not against the use of all external inputs in agriculture. Far from it! What we are saying is that farms have tremendous regenerative capacity that can be expanded and put to good use as farmers learn to use their internal resources more effectively.

Yes, your farm is worth more than ever. The title of this article is not an exaggeration. But for many farmers—especially those now following the high-input approach—the tremendous worth of their farms will remain obscured until they are able to base some of their agricultural efforts primarily on the tremendously powerful internal resources of their farms.

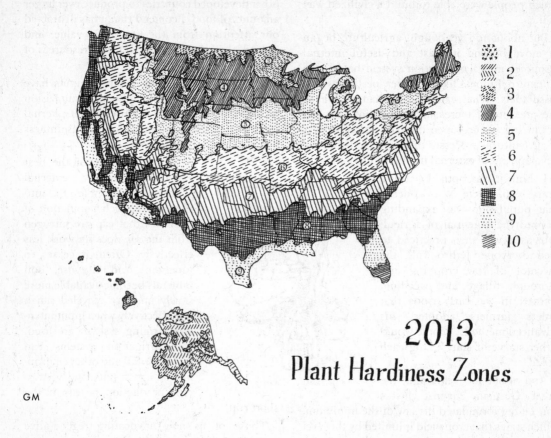

GM

2013
Plant Hardiness Zones

"HUMANS ARE A BIO-CHEMICAL
PHOTOGRAPH OF THE SOIL
THAT FEEDS THEM... —JOHN IKERD

EDITORIAL NOTE: Plant Hardiness Zones, otherwise known as USDA Zones were invented as a way to communicate simply a multitude of factors affecting farming and gardening, plant hardiness, growing days, hard frost, etc. Whatever the federal position on climate change, these zones have been changed continually over the past 50 years, and every farmer and gardener alive today can confirm for you without a doubt, that their zones are changing.

Early students of the American agricultural climates, soil types and growing regions were fascinated with the idea that you could predict not only what kind of plants would thrive based on climate and mineral composition, but also the relative vitality of the people nourished on that food, the quality of the horses raised on those meadows. We remember still today how Kentucky Bluegrass is supposed to grow the best horseflesh in the country. We remember less how the leached soils of Tennessee were destined to produce poor intellect and sickly dispositions in those who ate and lived there. -SvTF

FOOTNOTE: *Plant Hardiness Zones are cited from 2012 and have been have updated 3 times due to shifting conditions for gardeners and farmers.*

WILLIAM A. ALBRECHT

From William A. Albrecht
PANTHEONEER

Perhaps when we give ear more attentively to the voice of the soil reporting the ecological array of the different life forms (microbes, plants, animals, and man) over the earth, we shall recognize the soil fertility-according to the rocks and resulting soil in their climatic settings—playing the major role through food quality rather than quantity, surpluses, prices etc., in determining those respective arrays. Man still refuses to see himself under control by the soil. Then when we take national inventory of the reserves of fertility in the rocks and minerals mobilized so slowly into agricultural production in contrast to the speed of exploitation of virgin and remaining soil fertility, perhaps thrift and conservatism with reference to the mineral resources of our soils will be the first and foremost feature of our agricultural policy. By that policy we shall build the best national defense, if there is any basis for contending that if wars must come, food will win them and write the peace. May rocks their silence break and speak nationally through a better knowledge of soil for food as the basis of national health and thereby a national strength for the prevention of war and for the simplest road to peace. Our future national strength must rest in our soils.

FROM THE ALBRECHT PAPERS VOLUME I

EDITORIAL NOTE:
We have chosen William A. Albrecht as one of our pantheoneers for this Almanac because he was a long suffering visionary, considered a quack by many of his contemporaries— and forgotten by many avid students of organic farm history. He was a long outspoken agronomist working on cation exchange rations, observing fertility as reflected in animal health. With the advent of chemical fertilizers his findings were questioned by the university, he prevailed through force of character and left his papers to Charles Walters, founder of Acres USA - which exists still as a leading (and often radical) fount of alternative science for organic farming. -SvTF

From Lady Eve B. Balfour
PANTHEONEER

The most frequently heard argument is that intensive chemical farming provides the only hope of feeding the expanding world population and has therefore to be accepted whether we like it or not. To me it seems probable that the exact opposite could prove to be the case, and that it is an alternative and largely organic agriculture that will be forced upon us whether we like it or not. This is because, as is becoming increasingly apparent, the days of the former are numbered. One reason is the enormous demands on the world's non-renewable resources of energy, made by our Western life-style in general, and modern farming techniques in particular. Another is that modern methods are putting strains on the biota which is causing it to collapse.

Thus it is only common sense to look at alternatives, and in all seriousness study their potential viability.

It is not yet, however, generally accepted that the days of our present methods and behavior are numbered. Even where it is, it is too often regarded as a long-term problem which must not be allowed to obscure the immediate problem, namely the need to increase quantitative food production now. Here it is argued that organic farming is less efficient, that it has to rely on recycling which is wasteful, so that were it to be adopted, world food production would inevitably be lower, particularly production of protein, at a time when what we need is to produce ever more per acre.

To this I would like to point out three things:

1. A common view among nutritionists today is that the amount of protein (especially animal protein) hitherto thought to be required by man has been greatly over-estimated. (Organic farmers have found this also to be true for livestock).
2. There need be little loss in recycling if we did not waste so much.
3. Certainly we need to produce more per acre. Unfortunately the yardstick of modern economics is to measure the efficiency by production per man.

Labour-intensive small units will always be able to produce spectacularly more per acre than the large mechanized farms, apart from the finding that organically grown food goes further. When the inevitable change in life-style takes place I predict that we shall find it easier to feed the world population than we think, perhaps easier than now because Western nations will presumably have become less gluttonous. I predict also that we shall all be healthier!

FROM A 1977 SPEECH SHE GAVE TO THE INTERNATIONAL FEDERATION OF ORGANIC MOVEMENTS, IFOAM, WHICH HAS ITS ANNUAL CONFERENCE NEXT YEAR IN ISTANBUL, TURKEY. WWW.IFOAM.ORG
WWW.OWC2014.ORG

There will be a young farmers Delegation attending from the USA. Interested to join? Email farmer@thegreenhorns.net

The Living Soil

A BOOK WE RECOMMEND BY LADY EVE B. BALFOUR, FABER & FABER, LONDON, 1943

Eve Balfour was a key figure in the forming of the organic gardening and farming movement, and one of the founders of Britain's Soil Association. She divided her estate at Haughley into two sections, one organic, and the other run "conventionally", with chemical fertilizers. This book records the results, and her insights. "The health of soil, plant, animal and man is one and indivisible," she said. She practic ed what she preached, and reaped the benefits.

Lady Eve died in her 90s, healthy and alert to the end. This is an informative, thoughtful and inspiring book any gardener will want to read. Full text online at the Soil and Health Library. - SvTF

MORE INFORMATION: WWW.SOILANDHEALTH.ORG

GM

"There seem to be but three ways for a nation to acquire wealth. The first is by war... This is robbery. The second by commerce, which is generally cheating. The third by agriculture, the only honest way, wherein man receives a real increase of the seed thrown into the ground, in a kind of continual miracle, wrought by the hand of God in his favour, as a reward for his innocent life and his virtuous industry."

—Benjamin Franklin

NEW LIFE

Now I return to woods that I once knew
And I lie here upon the leaf-strewn ground
With ear attuned to hear roots breaking through
The unadorned and winter-lifeless ground . . .
For I do know when earth drinks rain that soon
New life stirs in the ever restless roots
Of violet, May-apple and percoon,
And ears can hear their tender bursting shoots
As they reach upward for the silver light
Of blowing wind, of sun and moon and star . . .
I know their stems seek life beyond the night
And nether world where worm and silence are.
And when I hear their breaking clods apart,
God, flower, stem, and dirt excite my brain;
I hear their little sounds above my heart . . .
Great growth to make the earth rejoice again.

- Jesse Stuart FROM KENTUCKY IS MY LAND

Now the natural man has a lot to do with nature. He keeps pet and cattle. He farms the land, and sometimes has strange feelings of awe and peace in the face of the power and beauty of the natural order. Indeed he frequently visits the parks rather than the pews, and thinks he is nearer to God's heart in a garden than anywhere else on earth including the communion rail.

What he does not know is that he is the priest of nature, that God has appointed him not only nature's crown, but nature's spokesman. He is to make the unceasing praise of nature articulate. The lark may not know that he is praising God in his melodious rapture, but man knows. Sometimes we suspect that the lark knows too, but not as man knows.

- David Head FROM HE SENT LEANNESS

VEGETABLE FARMER

URBAN FARMER

VETERINARIAN

BEEKEEPER

SOIL SCIENTIST

CATTLE RANCHER

BAKER

RIDING INSTRUCTOR

PRODUCE MANAGER

CHEF

BUTCHER

SEASONAL FARM WORKER

FARM CAREERS *by* GINNY MAKI

From the Northeast Beginning Farmers Project,
a BFRDP grant funded project, available as a
poster and a downloadable guidebook for
highschoolers and teachers.

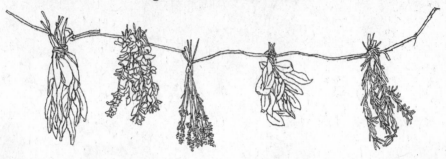

Table of Contents

• Apprenticeships & Farmer Ag Education
• Access to Land
• Capitalization of Small Diversified Operations: Business Planning and Accounting, Loans, Grants
• Getting started: What to grow, Buying Seeds, Greenhouses, Sustainable Pest Management, Soil Fertility, Irrigation, Wildlife & Habitat Preservation, Livestock, Equipment, Draft Animals, Staying Current, Laws & Regulations
• Models of Sustainable Agriculture: Urban Farming, Biodynamic Agriculture, Permaculture
• Marketing: Certifications, Farmers Markets, Retail, Value-Added Products, CSAs, Agritourism
• Community
• Farming Info clearinghouses: Government-Funded Programs & Resources, University & Extension Programs, Non-Profits, List Serves, Regional Networks, etc.
• Bio-bling: DIY projects, Yurts and Permaculture
• Big Picture: Research Watchdogs, Activism

MORE INFORMATION: WWW.NEBEGINNINGFARMERS.ORG
DOWNLOAD THE PDF HERE: HTTP://WWW.THEGREENHORNS.NET/WP-CONTENT/FILES_MF/1352147650GREENHORNSFIELDGUIDE2012LOWRES3.PDF

A NOTE
Regarding the Almanac

by NICHOLAS DANFORTH

Dear Greenhorns,

I can't resist writing to you, because you referred in your blog to our "forefathers", Ben Franklin and the others, who started the American almanac tradition. You're right that they were all men, but old Ben wasn't the first American to publish an almanac.

I think I'm the last direct descendant of Samuel Danforth, the poet who wrote the first almanac printed in New England, and one of the first tutors at the new college called Harvard. He was a popular poet and teacher, maybe in part because our family's apples, growing near what's now Harvard Square, were banned by the college when Harvard students drank too much apple-jack.

Great-great (x13) grandpa Sam's almanacs were full of agricultural and seasonal references, of course, and some of his verses had thinly-veiled political references questioning (British) government authority, an early and provocative tradition which I expect your almanac will continue – but not as veiled, I hope!

I like all you Greenhorns. I'm a young farmer (only 70), and I'm just retiring after 50 years of work (in the US and Africa). My farm, in the suburbs of Boston, is managed by my sisters (with advice and encouragement from my friend-for-life, Severine); it's about ten irrigated acres of terrific organic vegetables plus some farm animals. The work is done mostly by women. Because women globally have always done most farming and food-related work – judging from your blog, nowadays Greenhorns too are mostly women – maybe you should break with Great-Grandpa's tradition and call yours a "Farmer's All-WOMAN-ac" !

–Nick

EDITORIAL NOTE: Though the editorial team of this almanac is majority women, we have decided against this name change. - SvTF

FARMERS ARE GOOD CITIZENS.—It is universally admitted that farmers are the back-bone, nerve and muscle of our country, and it appears very strange that so many smart, industrious young men should hang about cities, and labor for others, year after year, for a mere pittance, while the vast fruitful field of the West lies open before them, and with a little exertion they can be lords of their own domain, perfectly free from the control and dictation of others. Let us suggest to them that to avoid the danger of going alone into a new country, they form companies of twenty, forty, or more, select a favorable locality in some of the Western States, and emigrate at once, feeling perfectly assured that they will bless the day when they resolved to become farmers. What is said on this subject, is said from experience, by one who was once a resident in the city, laboring for his bread, and is therefore now free to assure all who are similarly situated, that there is in the country more enjoyment of life, more freedom from want and care, more truth, religion and morality.   AGRICOLA.

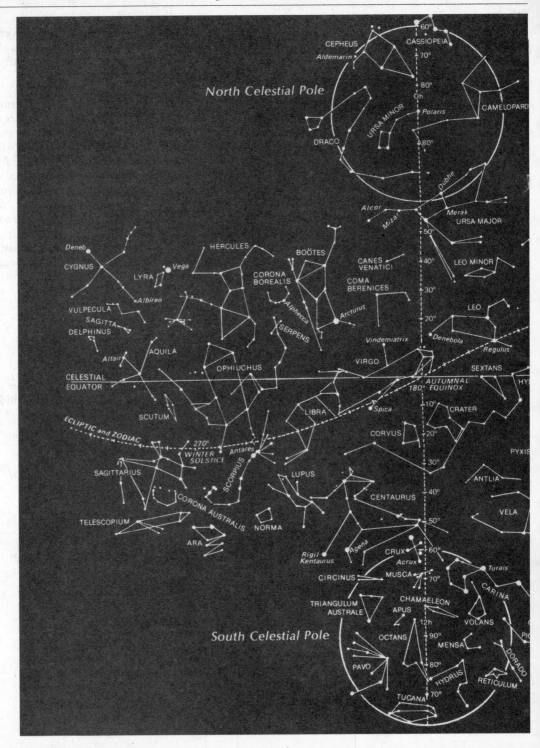

THE CONSTELLATIONS
AND THEIR BRIGHTEST STARS

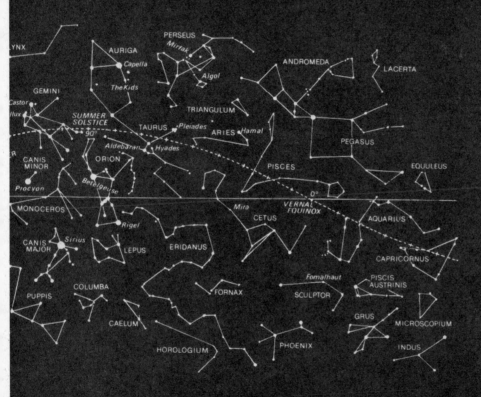

This present chart is an attempt to render the Starry World for our times. It is still based primarily on Ptolemy's descriptions. His catalogue, the Arabic star names, and other records have been consulted to accurately locate an "arm" or an "elbow", etc. Increasing familiarity with the constellations will make the use of the connecting lines unnecessary. Perhaps then the stars will be free to speak for themselves.

by STELLA NATURA - WWW.STELLANATURA.COM

Poem Beginning with a Line by Dana Ward

by ANNA MOSCHAVAKIS

GM

I was born on an earth much smaller
than the earth they're from.

I was born much smaller than they
were born on the earth they're from.

I was born cured in a similar warmth,
splendid and diminished, on an earth

much smaller than the earth
where they were born. I was born

diminished, lungs opened to the maximum,
opened to the material

ash, inert, that flows
from the earth where they are

The Moon.—Take care to procure good seed for all the crops you intend to raise, and have it sown early, have your grounds well prepared, and let them be well tended, and regard the *Moon* as *much*, and as *little*, as you please.

from. I was born
to effortless joy, much

smaller than the number of dead
buying rocks from the earth

they're from. I
was born neglected

in the maximum shark-light
in the dystopian wing

of the smaller earth I'm from.
I am not from the earth

they're from. They're from
another earth. They were born

on another earth. They are
not from the earth

I'm from.

Lord, I've been active all my life. This idea of eternal rest frightens me. The beatific something-or-other they talk about in sermons doesn't mean a thing to me. I shall be thoroughly miserable, if all I have to do is gaze and gaze. Isn't there anything to do in heaven?

- David Head

FROM HE SENT LEANNESS

New Blood For The Old Body

BY TRACE RAMSEY

Many of us never meant to become farmers. We had our ambitions to enter the world as accountants or lawyers or teachers or some other clean, respectable professional. We never really thought about the origins of our food or questioned the intentions of those who screen out the realities of farming; we always knew that the supermarket shelves would fill themselves, food came in boxes or cans ready to serve, and farmers were simply one dimensional photographs in the mix of a hot new marketing campaign. Sustainable and industrial agriculture held meaningless differences, no more distinction than competing national brands of light duty trucks or diet soda. But then something happened. In the previously steady route of our lives, a shift occurred. The soil moved under us somehow, got stuck in the creases of our pants, in the ridges of our shoes, in the lines of our palms. Suddenly white picket fences, situation comedies and mutual fund returns didn't seem so interesting anymore. The big ball game and the driving range became distractions from the reality of a new love affair. We got hooked on the possibilities of growing our own food and also providing that food to others.

The epiphany was likely different for many of

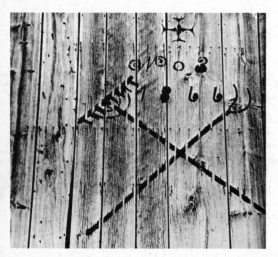

us. Maybe a friend took us to a farmers market. Maybe someone had a plate of local hamburgers or collards at a picnic. Maybe the news of some global food disaster made us question the monocultures piled high on our plates. Maybe a real life farmer entered our life.

For a few of us, those with farming in our past – a childhood spent in the fields of the big farms or the family plots, throwing rocks into the hedgerows for little or no pay or watching over milking machines in the stench of industrial-sized barns – there was no love, no kind of encouragement, no appreciation for our part in the dynamics of food production. We were simply limbs and calluses then, small gears in a giant cranking clock. We left the farm to pursue something else only to be pulled back hard when it became apparent that we could abandon everything that farming once meant to us. We could make it ours.

Still others came to farming from DIY and anti-authoritarian backgrounds, building urban community gardens or putting up food in anarchist collectives. Gardening always had a community aspect to it, but we wanted something more. We knew that we could do the work, that we had the right vision and skills. We just needed the access and the resources to get started. Regardless of how we arrived at this point, here we are; we will call ourselves farmers from now on. We are transplants from cities, dropouts from university systems and ex-corporate shufflers. We are mothers and sons and grandparents, masters in communications, colorful documentarians, shy propagandists. Most of all, we are teachers and students inhabiting the same bodies and breathing the same air.

Our young and new farmer movement is made up of many itinerant folks, traveling to places we want to see, gaining knowledge we never thought we would need. Our commonality with the landed and the stable is the soil and its layers. More specifically, our bond is in the ways we approach that soil and our desire to grow food in a way that builds on a sense of the farmer never dying. The immortality is not functional but symbolic – if you imagine that you will need to use a piece of soil forever, you will never intentionally do it harm.

This intentionality is not a new idea, but neither is it very well known in the information age. It is buried in our collective past, not necessarily waiting to be discovered, but intact and beckoning nonetheless. To get to the guts of it we are throwing away the agricultural methods of our parents and grandparents, even subverting our great-grandparents' proud thoughts of survival amidst the coming surpluses. Things may appear as cobbled together bits of dust and weight and worn out shovels, but its functionality in an agrarian way of life is apparent with very little inspection.

We stand in the books and plots and ideas of the past, pulling out the rusty pages and diseased cells in order to build something practical from the obsolete and misinterpreted, rewiring the seed catalogs, rewilding the crosswalks, reconnecting the pastures to the kitchens.

So here we are, doing more than is required of us, daily pushing the boundaries of our bedtimes, our muscle structure, our hunger pains, our balance of minimalist living conditions with the reality of satisfying relationships. We don't need justification for living this life, but that rejection of validation won't feed or shelter our families

or protect our chickens from roaming dogs. We have concrete needs – access to land, to capital, to markets – but we cannot ignore the bounty before us as we seek to satisfy these.

We have to live farming as it happens, at our level, at the pace that we can move. The weeds don't and won't pull themselves; the new beds won't magically appear out of spilt potting mix or the crumbs of a quick dinner of sandwiches among the paths. Anyone who tells you that growing food is simple is a lunatic. Anyone who tells you that having animals lessens the physical workload is a liar. But we stick the possibilities of a simpler, easier way of life in the context of the larger ecology, the massive inebriation that defines the world and my generation. If we are to sober up, we need to get moving.

We are bridging eras, going about tasks the hard way but with newer tools and even newer outlets, burrowing into ancient methods and supplementing with our own big-brained flourishes. A generation of reclamation, telling our story to groups of people that may have never been inspired to so much as think about how a piece of grass might pop from a crack in the sidewalk. The whisper is that we are here to exploit those cracks, get our dirty fingernails scratched with asphalt and debris while attempting to save the disorientated souls of the material apocalypse.

STEADINESS OF PURPOSE.—In whatever you engage, pursue it with a steadiness of purpose, as though you were determined to succeed. A vacillating mind never accomplished anything worth naming. There is nothing like a fixed, steady aim. It dignifies your nature, and insures your success. Who have done the most for mankind? Who have secured the rarest honors? Who have raised themselves from poverty to riches? Those who were steady in their purpose. They move noiselessly along, and yet what wonders they accomplish! They rise—gradually, we grant —but surely. The heavens are not too high for them, neither are the stars beyond their reach. How worthy of imitation!

We young farmers have the double task of growing food for the community as well as being able to communicate about the process and our decisions in spaces that are new and possibly uncomfortable.

The pictures we take of ourselves hang in art shows and stand out in glossy magazines; our recipes are printed on cardstock and handed out at tradeshows; our words bring excitement to readers wishing that they too could participate in the riot that is small-scale sustainable agriculture. This riot exists outside the handshakes and millionaires of the agra-political grease machines, knowing, with the certainty of the tides, that the transactions we despise will occur no matter how long we scream, no matter how far we march, no matter how many letters we write. It is not defeatist or abandonment of the successful tactics of the past, just recognition that we can do much better with the actual actions of farming in sustainable ways, demonstrating to the consumers and wholesalers and value-adders that we are successful despite their dismissals. We cannot change the culture without changing the culture; yelling and otherwise carrying-on never has set a sweet fruit or fed a piglet, and I bet it never will.

We love this life – we have to – but sometimes we can feel that we don't own it, that it owns us and grips us in a way that will never shake us loose. In those moments of weight we can only shrug, pull on the rubber boots and move deliberately until the fireflies speckle the whippoorwills' breaths. Throughout all the highs and lows we can look at ourselves over and over again knowing that, if we stick to our ideals, we can do noble and appropriate work no matter what happens.

We are the new blood in the old body.

FROSTS AND FREEZING WEATHER

IT might be assumed that frosts and freezing weather are of interest only in the more northerly parts of the United States. As a matter of fact, freezing temperatures have occurred at some time in every section of the country with the exception of the Florida Keys. In normal years all parts of the country except southern Florida, the Gulf coast, southern California and certain favored areas in Arizona experience frost. The lowest temperature on record in the United States is sixty-six degrees below zero (Fahrenheit). It was recorded at Riverside, in the Yellowstone National Park, Wyoming.

A Great-Grandpa for Every Young Farmer: Mr. Eric Sloane

by SARAJANE SNYDER

It seems obvious to say that our generation of farmers is seriously lacking in intergenerational know-how.

Of course there are those among us who inherited skills, knowledge, tools, and land—all of which have their own priceless value (and, oftentimes, complications). For many of us, though, born roughly between 1980 & 1990, our parents (or in some cases our grandparents) worked hard to get away from farming. Either that or the family tree is completely lacking in farmers altogether.

We young farmers are left with so much to learn that it can be sometimes overwhelming. We haven't had the lifelong immersion in the communities of our great grandparents, be they in Ireland or Australia, Cambodia or Connecticut, where small-scale agriculture was a given fact, where people necessarily lived within their means, and communities of people were held together by shared work and tradition. Very little of the agrarian lifestyle is a given anymore.

There are many ways to tell the story of how we

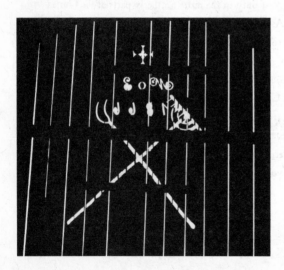

got from then to now, and many of those versions highlight successes, great gains, triumph. But every honest telling of the story must include some recognition of the great losses we have suffered.

Eric Sloane (1905-1985) was a prolific artist and amateur historian with a keen, well-researched, and heartfelt sense of what we've lost in the passing generations. His nearly 40 books lovingly document practical hardware & structures like barns, fences, and tools of house and farm, but they also breathe life back into forgotten customs, traditions, and skills.

Sloane's work is an amazing resource of old ways, often dating back to the Colonial era. The design and production of his books, creamy heavyweight pages filled with hand-lettering, line drawings and little captions and asides, brings life to "the early Americans" for whom Sloane has utter admiration. He reminds the reader that our country was founded on agrarian ideals by people who knew how to work hard and think practically. He also, with intention, illustrates how rich and varied life was before the industrial revolution. He creates a world where there is no noise pollution, no light pollution, and a deeper appreciation of quality, of resources, and of small gifts.

Sloane's own personality, which comes across as fairly grandfatherly, is an important part of his work. When you read a book of his, you feel the warmth of his gift being given to you.

In addition to books about tools and early American customs, he also has written numerous books on the sky, clouds, and weather forecasting, which are essential reading for anyone who works outside under the sky. Sloane's book Folklore of American Weather works through a treasury of American folk sayings around weather, stamping them True, False, or Possible. His other major weather book, Look At the Sky and Tell the Weather, gives a tutorial in just what the title promises [note, I would like to give a better description of this book, but I have to find it and read it first! It's Sloane's favorite of all his books...] Observing and tracking the weather,

learning to read it and listen to the signs and be your own forecaster is a great way to improve your sense of belonging to a place. Eric Sloane knows that, and he, in all his works, encourages people to belong to their place.

In his book The Spirits of '76, Sloane presents ten "spirits" that he considers absolutely essential for our forefathers (the book was written at the bicentennial in 1976), but which have faded almost entirely from our current culture.   His list of ten includes respect, hard work, frugality, thankfulness, pioneering, godliness, agronomy, time, independence, and awareness.

"The Spirit of Agronomy" is one of the spirits that is most important to our continued health and vigor, and yet so delicately held. Unlike, say, thankfulness or hard work, the spirit of agronomy is not exactly an abstract principle of which we can each summon our own version. Agronomy is a spirit–it's a spirit of respect for what a farm organism is and has to offer, it's a spirit of understanding the subtleties of working with the land, it's a spirit of self-sufficiency, & long-refined knowledge, but it is a spirit that moves within a craft, a discipline, a skill set. It is a spirit that seems to be dying more fully with each generation of elders that leaves us. Even though there are more and more young people learning the ways of small-scale farming, re-instilling the spirit of agronomy, it is something that truly only can mature through generations. Because he inspires us to remember the past and revisit the old skills, Eric Sloane's legacy to young farmers is and will continue to be invaluable.

The Matter of Light

by ROSMARIE WALDROP

A SWALLOW CUTS an arc along the roofs, cuts it again, as if to move the horizon inward. Light spills through my chest, stirring up armies of pigment. The gulls cry like babies, and clouds cut the distance with splendid, unnecessary profusion. The movements that my body is afraid of, your shirt on the line does in perfect harmony with the elements.

The depth of a river is measured in drowned bodies, but the laws of nature copulate madly, in moth frenzy. I remember the pines and poplars, their reflection, too, drowned, and the white handkerchief waving and waving to cut the distance. I meant to tell my troubles and sail on to a wholly new identity. This curve was headed for the high turbulence of moist dreams.

The years we've lived together are piled on the path to the shore. In splendid profusion. We can walk on them now, with the air cooling. Turning a "No" mask slightly downward is known as "clouding" because the mask takes on a melancholy aspect. The way my eyes had the run of the sky, but were defeated by its blinding excess. Bands of white foam ripple out into the Bay where the river comes undone. The contour of the rocks, like the mask, is meant to be looked at from a distance, yet is most alive close up, in light rain.

We had leaned out the window and found night early. Hugged the bed in alarm and dived below the skin. To move the horizon inward, make clouds drift through our bones. The swallows had drowned in their reflection, along with the pines and poplars. And the handker¬chief waving and waving. If there is enough deep red in the landscape will even an old woman's embrace spill light?

AH

Violence broke out in Loup City Nebraska when Communists tried to support a strike in a chicken plant.

Radical Farm Protests

by BILL GANZEL OF THE GANZEL GROUP
FIRST WRITTEN AND PUBLISHED IN 2003

As the Great Depression worsened, farmers across the country began looking for new and sometimes radical solutions to their problems. As early as 1932, some farmers were trying to raise agricultural prices by physically keeping produce off of the market, on the theory that if they could reduce the supply, both demand and price would rise in response. Some of the most radical events took place in Nebraska, culminating in the violence at Loup City.

In Iowa and Nebraska, a group known as the Farm Holiday movement built road blocks on the highways leading to the agricultural markets. Although they dumped some milk and turned back some cattle, the blockades weren't effective, and police eventually opened the roads. In February 1933, thousands of farmers marched on the new capitol building in Lincoln demanding a moratorium on all farm foreclosures. The legislature did pass a two year moratorium, but it left in a loophole, which allowed judges to order foreclosures to continue at their discretion.

As the Depression continued, some in Madison County began listening to a fiery Communist organizer, "Mother" Ella Reeve Bloor. Mother Bloor had come to the Midwest to build alliances between urban workers and radical farmers. Throughout 1933 and '34 she spoke all over Nebraska, attempting to recruit the disaffected and dispossessed to her cause.

In Loup City there were two clearly defined factions, each with a newspaper telling its side of the story. One newspaper, The Standard, demanded higher farm prices, cancellation on payment of feed and seed loans, a moratorium on mortgages and reduced taxes. The other paper, The Times, called for the "American Legion boys of Sherman County" to become vigilantes and not "allow a communist to come to Loup City, speak from a platform in the district courtroom, openly insult every Legion boy by calling them 'cowards' because they dared fight for their country's flag."

Despite the rising tensions, Mother Bloor often spoke in Loup city, sometimes accompanied by an African American couple from Grand Island, Mr. and Mrs. Floyd Booth, who were also Communist organizers.

In June, 1934, the government designated Loup City and Sherman County as among the worst drought areas of the plains. That same month, young women who were working as chicken pluckers in a local creamery plant were talking about striking for better pay. Mother Bloor, the Booths, and others from Grand Island announced they would travel to Loup City and speak in support of the strike.

When they arrived on June 14th, Flag Day, they spoke in the town square, and then marched to the factory. There the managers of the plant gave in to some of the demands, but refused to recognize any kind of union representation. The radicals marched back to the town square.

They were followed. A group of locals and guards from the plant confronted the farmers. Someone yelled, "Hey Rube" which apparently a signal. Suddenly, the two groups were fighting each other. Fists flew. Blackjacks made of bars of soap in stockings knocked several people unconscious. Some were carried to the hospital. Others fled in their cars and trucks. The rally was over.

The Times rejoiced that "Red blooded citizens in Sherman county displayed their loyalty to the Stars and Stripes last Thursday when they drove 'Red' invaders who came here looking for trouble out of Loup City." The article suggests racism both on the part of its author and of the strike breakers; "there are not over a dozen farmers in Sherman County who are in favor of importing a colored man from Grand Island [Booth] or anywhere else to stir up trouble for them."

• •

EDITORIAL NOTE: A penny auction is where farmers attending bank foreclosures at a farm bid only one penny for each item, in this way preventing the bank from liquidating the farm's assets, and allowing the farmers another shot at solvency. This was economic solidarity commonly shown in this era of foreclosures. I have not read about such a practice in our current housing crisis. - SvTF

Fighting Farmers

PHOTOS FROM GOOGLE IMAGES

Recent European protest against EU agricultural policies.

NOVEMBER 2012
Tractorcade

Direct Action, Direct Market in Four Movements

by AMY FRANCESCHINI, FUTUREFARMERS

SUPERPATAT, FIELD LIBERATION MOVEMENT, BELGIUM, 2011

Over the course of this year a number of actions and movements have emerged. Demands by farmers for a more direct exchange with their consumers (Direct Market) combined with direct action has given rise to inspiring tactics and aesthetic actions.

If we are to demand a better world we must also imagine what this might look like. If we do not demonstrate the beauty and justice we wish to see, how will they know what to look for? The role of visual and performed expression is to proliferate- to go beyond the symbolic and penetrate our everyday life. To enliven the actions with tools and performed display of our dreams. More reasons for farmers and artists to collaborate! (oh yes, and to make more films.)

1. Erechim, Sarandi and São Lourenço do Sul

In Erechim, hundreds of farmers gathered in front of the Banco do Brazil in September Seventh Avenue, a major city. They blocked the passage of vehicles and prevented customers from entering the bank. During the morning, the producers made threshing corn and bags of the product delivered to the bank manager, as a symbolic payment of debts.

THE FIELD LIBERATION MOVEMENT WETTERN, BELGIUM, MAY 29, 2011

Hundreds of activists freed a GM potato open test field at the university of Ghent in Wetteren. They pulled out genetically modified potatoes and swapped them with organic potatoes. The "Great Potato Swap", was an openly announced public and non-violent event. Demonstrators came with their chairs to assemble in the streets for meetings, the haunting "Superpatat" sculpture made of a giant black fabric covering 5 people that would amble down the road and a potato canon to fire organic potatoes into the liberated field. The action stimulated debates in the papers, online, and on the radio and television about GMO's.

What struck me about this action and movement was their preparation tactics and transparency. They told the local police they were coming and met in person with them in the weeks prior to the action.

A few months before (Sept. 16, 2010) the field liberation, they instigated a training session on the lawn in front of the European Parliament in Brussels, Belgium. Dressed in hasmat suits, they planted several potato plants and corn plants and taught people how to cut them down and pull them out with speed and grace. They were completely transparent about what they were training for with the press and passerby.

Presently (Dec. 2012) 11 activists, now called "Potato Eleven" have been accused for belonging to a criminal gang, and risk being asked to pay 200,000 euros in damages.

The next meeting in court will take place January 15th where the main discussion and argumentation will take place. The lawyers worked out in collaboration with FLM three main tracks of defense for the activists:

- the right to demonstrate and to express opinion.

- necessity for action. the permission by the biosecurity council for the test field wasn't completely done according to the rules of the procedure (conflict of interests cause many professors implied in the test field are also part of the security council).

- the denial of the accusations by the university of Ghent. there was no criminal organisation. there was no use of violence by activists. activists can't be masde accountable for the decision of the university to invest in heavy security for 6 months.

COUNTRY ENTERS THE CITY
RIGA, LATVIA
MAY 22, 2012

A symbolic activity of Latvian townsmen and farmers – "Country Enters the City" – was launched as a result of cooperation of the three largest farmers' organisations – Latvian Agricultural Cooperatives Association (LLKA), Farmers Parliament and Agricultural Organisations Cooperation Council (LOSP). Instead of the traditional greenery with flowers at the Cabinet of Ministers which is the central location in both symbolic and literal sense, this year the bed was filled with edible flora. Cabbage, beetroots, lettuce, beans and spices flourished along with common rural flowers such as marigolds, French marigolds and nasturtiums.

This campaign demonstrated a symbolic alliance of townsmen, public administration officers and politicians with the farmers in their fight for equal direct payments. In the creation of the vegetable bed the farmers are supported by State Chancellery.

Latvia and the other Baltic States are not satisfied with the Common Agricultural Policy (CAP) reform after 2013, proposed by the European Commission in October last year (KLP). Regardless of the fact that the government and the responsible officials have defined the farmers' interests as a European priority issue, the farmers have decided to get involved in the protection of their interests by launching an EU direct payment lobbying and public relations campaign. Considering that Latvia is a small State with a restricted representation at the European level, one of the aims of the campaign is to involve a maximum variety of people by bringing together the townsmen and farmers. The vegetables bed at the Cabinet of Ministers is one of the activities of this campaign.

Vegetable plants for the beds are provided by the cooperative Mūsmāju dārzeņi. The soil is prepared by the Earthworm Growers' Association and Baiba Rudzāte, landscaping gardener.

THE POTATO MOVEMENT
NEVROKOPI, GREECE
FEBRUARY 5, 2012

Free distribution of more than ten tons of potatoes in the center of Thessaloniki, was organised by a group of farmers from the village of Nevrokopi, Greece's potato-growing capital.

The farmers were protesting against imports of Egyptian potatoes - while they had barns full of the Greek product - after a meeting between the agriculture minister and potato importers days earlier failed to yield any concessions.

The first to invite the farmers of Nevrokopi to sell their potatoes at 25 cents a kilo - one-third of market price, was the Pieria Prefecture Voluntary Action Group, based in the northern Greek town of Katerini.

Keratsini Mayor Tzanis: 'We did not come to hold sway over the city, but to serve the city'. The Pieria group was formed in late 2007, after a series of wildfires devastated Greece's forests, to provide the local fire service with an early warning system. It was already busy creating a free supermarket for the destitute when it heard of the potato handout in Thessaloniki.

Days later, the movement spread to Thessaloniki's Aristotelian University. Christos Kamenidis, a professor of agricultural marketing, organised a potato sale on campus with student volunteers. "I was worried that we would [only] sell three or four tonnes. We sold 50 tons on the first day".

The potato movement is changing the food market. The stores that once sold for 75 cents are selling for 45 cents. And the effect has not been limited to potatoes. Other basic durable goods such as olive oil, flour, rice, and honey have also gone on sale directly from producers, undercutting market prices by half.

How it works:
A town hall announces a sale. Locals sign up for what they want to buy. The town hall then tells Kamenides the quantity required and he and his students call local farmers to see who can supply it. They show up with the requisite amount of produce at the appointed place and time, meet their consumers, and the deal is done.
The direct sales are immensely popular. One organised last month by volunteers in Katerini, south of Thessaloniki, last month saw an online offer of 24 tonnes of potatoes sell out within four days, with 534 families pre-ordering.

Postscript:

Why more art? Why more films?:

"Politics (from Greek politikos "of, for, or relating to citizens") is the art or science of running governmental or state affairs, including actions within civil governments, but also institutions, fields, and special interest groups such as the corporate, academic, and religious segments of society."

If we focus on the "of or relating to citizens" part withing the definition of politics, then I am reminded of two letters that were sent to Russian film maker, Andrey Tarkovsky about a film he made called Mirror.

One was from a worker in a Leningrad factory. He said, "It is a great virtue to be able to listen and understand... That is a first principle of human relationships: the capacity to understand and forgive people their unintentional faults, their natural failures. If two people have been able to experience the same thing even once, they will be able to understand each other. Even if one lived through the era of the mammoth and the other in the age of electricity."

This is what art can do- help us, even for a moment, understand each other.

Another letter he received was from a woman whose daughter had written her a letter about Mirror:
"How many words does a person know? How many do we use in our everyday vocabulary? One hundred, two, three? We wrap our feelings up in words, try to express joy and sorry and any sort of emotion, the very things that can't in fact be expressed. But, there is another kind of language, another form of communication; by means of feeling and images. That is the contact that stops people being separated from each other, that brings down barriers. Will, feeling, emotion– these remove obstacles from between people who otherwise stand on opposite sides of a mirror. The frame of the screen moves out, and the world which used to be partitioned off comes into us, becomes something real...And this does not happen through little Andrey, it's Tarkovsky himself addressing the audience directly, as they sit on the other side of the screen. "

I found these words incredibly moving and would only expand on her words a bit to include this screen she mentions in the streets, on farms, inside public institutions and to possibly work without a "screen" or gallery wall- to be in directly in the situation- not only representing it. I say this not to discount the power of film, but to illicit a way of working that includes the viewer in the production of the work. Tarkovsky does this using film, while, I wish to find a mode where an audience does not receive the work in a dark room, but in the light of day.

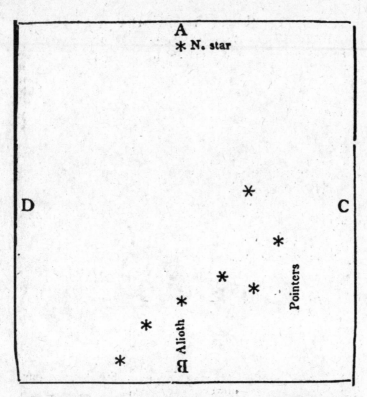

The above figure represents the North Star, Alioth, and the Pointers, when Alioth is on the meridian below the pole. By turning the figure upside down, it will represent them when Alioth is on the meridian above the pole (or North Star) and by turning the figure one quarter round, so as the letter C on the right hand of the figure appears to be overhead, or place the figure so that the letters C and D become perpendicular, then it will represent them when Alioth and the Pole, or North Star, are on a level east and west line, Alioth appearing east of the Pole Star. By turning it the other way, so that D is over head, it will represent them on a level east and west line, and Alioth will then appear west of the North Star. Any person, by viewing the above figure, and comparing it with the Stars themselves in the northern hemisphere, may become acquainted with and know the North Star, Alioth, and the Pointers.

N. B. The above Stars make a complete revolution round the Pole once in about twenty-four hours, and are called the Waggon, or Plough.

Of Moon Law

by ELEANOR EICHENBAUM

Underwater, we became blue
Sound here is thick and grace expected
Blue is warm
In warmth, we forgot standing and the medium of air
To swim is a feeling, like learning to walk
Sometimes nets of circles pulse in the sky
When it stops, we don't call it a storm
Everything moves here and we change
More grace

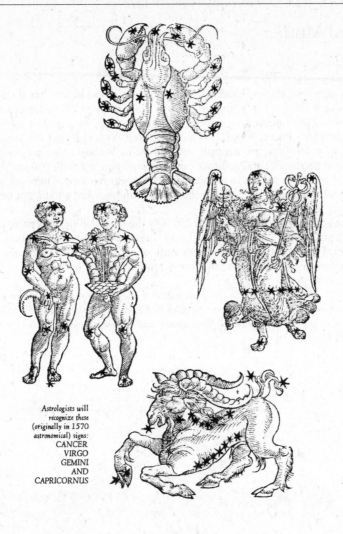

Astrologists will
recognize these
(originally in 1570
astronomical) signs:
CANCER
VIRGO
GEMINI
AND
CAPRICORNUS

♊ Gemini, Arms   ♉ Taurus, Neck

♋ Cancer, Breast   ♌ Leo, the heart

♍ Virgo, Bowels   ♎ Libra, Reins

♏ Scorpio, Secrets   ♐ Sagitarius, Thighs

♑ Capricorn, Knees   ♒ Aquarius, Legs

                                  Pisces ♓ the feet.

Albany Almanac, 1815

The Revolt of Mother

by MARY WILKINS FREEMAN

"Now, father, look here" Sarah Penn had not sat down; she stood before her husband in the humble fashion of a Scripture woman. "I'm going to talk real plain to you; I have never sence I married you, but I'm going' to now. I ain't never complained, an I ain't going' to complain now, but I'm going ' to talk plain. You see this room here, father; you look at it well. You see there ain't no carpet on the floor, an' you see the paper is all dirty, an' droppin' off the walls. We ain't had no new paper on it for ten years, an' then I put it on myself, and it didn't cost but nine-pence a roll. You see this room father, it's all the one I've had to work in an' eat in an' sit in sence we was married. There ain't another woman in the whole town whose husband ain't got half the means you have but what's got better..."

She stepped to another door and opened it. It led into the small, ill-lighted pantry. "Here," She said, is all the butter I've got - every place I've got for my dishes, to set away my victuals in, an' to keep my milk-pans in. Father, I've been talkin' care of the milk of six cows in this place, an' now you're going' to build a new barn, an' keep more cows, an' give me more to do in it."

The Revolt of Mother, short story by Mary Wilkins Freeman published Harpers 1890. Excerpted from ' Putting the Barn before the House" Women and Family Farming in early 20th Century New York, Grey Osterud. This book is a history of gender roles, rural institutions, the balance of farm-enterprises, off farm income, and a changing social matrix in one of the most prosperous farm regions of the North East.

Who owns most   of the land?

AML

"The Land question is nowhere a mere local question; it is a universal question. It involves the great problem of the distribution of wealth, which is everywhere forcing itself upon attention. It cannot be settled by measures which in their nature can have but a local application. It can be settled only by measures which in their nature will apply everywhere. It cannot be settled by half-way measures, it can be settled only by the acknowledgment of equal rights to land. Upon this basis it can be settled easily and permanently... And whether the Land Leagues move forward or slink back, the agitation must spread to this side of the Atlantic. The Republic, the true Republic is not yet here. But her birth struggle must soon begin. Already, with the hope of her, men's thoughts are stirring. Not a republic of landlords and peasants, not a republic of millionaires and tramps, not a republic in which some are masters and some serve. But a republic of equal citizens, where competition becomes cooperation, and the interdependence of all gives true independence to each, where moral progress goes hand in hand with intellectual progress, and material progress elevates and enfranchises even the poorest and weakest and lowliest."

- Henry George, 1881

FROM THE LAND QUESTION ETC.

EDITORIAL NOTE: The Land Question is mostly about the terms of tenant farming in Ireland compared to the Continent, but he also travels to America to make a comparison of 'freedom' of the farmers internationally, to understand the economic terms of tenantry.- SvTF

Frost my birthday a
Lavender cake, new binoculars
A cardinal in the field of sheep

–Mariette Lamson

FEBRUARY

—

THAW

FEBRUARY 2013

1

♍ F

♑ ☉ 7:06 17:14

---

2 Today is Candlemas: one of the great cross-quarter days which make up the wheel of the year. It falls midway between the winter solstice and the spring equinox and in many traditions is considered the beginning of spring.

♍ sA

♑ ☉ 7:05 17:15

---

3 This week, Mars will still be visible above the setting Sun.

♎ SU

♑ ☉ 7:04 17:16

4

♎ ♏ M

♑ ☉ 7:03 17:18

5

♏ TU

♑ ☉ 7:02 17:19

---

6 2013: Pennsylvania Association for Sustainable Agriculture annual Farming for the Future Conference begins (pasafarming.org)

♏ ♐ W

♑ ☉ 7:01 17:20

---

7 2013: Organicology Conference begins in Portland, Oregon (organicology.org/)

♐ TH

♑ ☉ 6:59 17:21

8

♐ ♑ F

♑ ☉ 6:58 17:23

---

9 in 1877 the US weather service was established and in 1891 the first shipment of asparagus arrives in San Francisco from Sacramento

♑ SA

♑ ☉ 6:57 17:24

---

10 When two planets occupy the same degree along the ecliptic (path through the celestial sphere), it is called a conjunction. On this day, the Sun and the Moon are in conjunction, resulting in a new moon.

NEW

♑ SU

♑ ☉ 6:56 17:25

11

♒ ♓ M

♑ ☉ 6:55 17:26

12

♓ TU

♑ ☉ 6:53 17:28

---

13 in 1837 there is a Flour Riot in New York City 6,000 workers at a "bread, meat, rent, & fuel" meeting in Chatham Square assault local flour merchants who, they claim, are hoarding flour in order to drive up the price.

♓ W

♑ ☉ 6:51 17:29

14

♓ TH

♑ ☉ 6:51 17:30

15

♓ ♈ F

♒ ☉ 6:50 17:31

---

16 The week of George Washington's birthday was designated as National Future Farmers of America Week in 1947. FFA Week gives members a chance to educate the public about agriculture.

♈ SA

♒ ☉ 6:48 17:32

17

♈ ♉ SU

♒ ☉ 6:47 17:34

---

18 in 1930, while studying photographs taken in January, Clyde Tombaugh discovers Pluto.

♉ M

♒ ☉ 6:46 17:35

---

19 Each zodiacal constellation is associated with one of the four elements (Fire, Earth, Air or Water), which in the Stella Natura calendar corresponds to different parts of the plant (Fruit, Root, Flower or Leaf). When the moon is in a certain constellation, growth in the corresponding part of the plant is stimulated.

AG

♉ TU

♒ ☉ 6:44 17:36

20

♉ ♊ W

♒ ☉ 6:43 17:37

---

21 In 1741, Jethro Tull, inventor of the Seed Drill dies at the age of 67

♊ TH

♒ ☉ 6:41 17:38

22

♊ ♋ F

♒ ☉ 6:40 17:39

23

♋ SA

♒ ☉ 6:39 17:41

24

♋ ♌ SU

♒ ☉ 6:37 17:42

25

♌ M

♒ ☉ 6:36 17:43

26

FULL

♌ ♍ TU

♒ ☉ 6:34 17:44

27

♍ W

♒ ☉ 6:33 17:45

---

28 in England in 1936 Emma Goldman gives a lecture to the Workers Circle in London.

♍ TH

♒ ☉ 6:31 17:46

---

zodiac signs

Cancer ♋ Good for planting, irrigating, grafting + transplanting. Most Fruitful	**Scorpio** ♏ Exceptional for vine-type growth. Fruitful	**Pisces** ♓ Good for root crops but a very poor time to make anything! Fruitful
Taurus ♉ A time to plant lettuce, cabbage, sturdy stalks + root crops. Semi-fruitful	**Capricorn** ♑ Good time to plant tubers + root crops. Semi-fruitful	**Libra** ♎ Time for seeding of hay, grain, flowers and crops where beauty is concerned Semi-fruitful
Aries ♈ Time for cultivating + plowing where weed destruction is sought after. Chickens hatched now will be robust Semi-barren	**Sagittarius** ♐ Best for harvesting, cultivating + onion sets. Semi-barren	**Aquarius** ♒ Suitable for harvesting + destroying of unwanted growth. Barren
Virgo ♍ Good for bringing order to the garden in general - pruning. Barren	**Gemini** ♊ Time to cultivate - destroying weeds. Barren	**Leo** ♌ A time for killing weeds + unwanted growth. Cider should be drawn off for storage. A good time to cut hay - if earth has to destroyed, cut back to the new of the moon. Most Barren

THE NEW OLD TRADITIONS
THROUGHOUT THE YEAR

"The Story of the Sun and Moon"

by ROBERT DOTO

The passing of time throughout the year may be understood as a story involving two main characters, the Moon and the Sun, who every year play out their comedic tragedy of life, death, and rebirth, providing a narrative and reason to gather for all people.

THE PUSH: "The Labor of Love" (2nd to last weekend in February)

For those of us living in New York City, the period between New Years and Spring is a time to battle against the cold. Even with the Sun burning brighter, the air still becomes more frigid. Change can be difficult, and the bitterness we feel can be attributed to the birth pangs of a new time of year emerging. The Moon learns how to carry her child, while her child learns how to emerge from her womb. Both want to be simultaneously with and separate from one another. The Push allows us to help each other through this time.

Under the double spell of night and frost
Within the yeoman's kitchen scheme
The year revolves its immemorial prose
He reckons labor, reckons too the cost
Mates up his beasts, and sees his calf-run teem...

-Vita Sackville West

MEDICAL USES OF SALT.

In many cases of disordered stomach, a teaspoon full of salt taken three times a day, is a certain cure. In violent internal aching, termed cholic, add a tablespoon full of salt to a pint of cold water, drink it and go to bed; it is one of the speediest remedies known. The same will revive a person who seems almost dead from a heavy fall, &c. In an apoplectic fit no time should be lost in pouring salt and water down the throat, if sufficient sensibility remains to allow swallowing; if not, the head must be sponged with cold water until the senses return, when salt and water will completely restore the patient from the lethargy. In the fit, the feet should be placed in warm water, with mustard added, and the legs briskly rubbed, all bandages removed from the neck, &c. and a cool apartment procured, if possible. In many cases of severe bleeding at the lungs, when other remedies fail, Dr. Rush found two teaspoons full of salt completely stayed the flow of blood. In cases of bite from a mad dog, wash the part with strong brine for an hour, then bind on some salt with a rag. In toothache, warm salt and water held to the part, and renewed two or three times, will relieve in most cases. If the gums be affected, wash the mouth with brine; if the teeth be covered with tartar, wash them twice a day with salt and water. In swelled neck, wash the part with brine, and drink it also twice a day, until cured. Salt will expel worms if used in the food in moderate degree, and aid digestion; but salt meat is injurious if much used.

EACH GALLON OF SEA WATER
CONTAINS
0.2547
POUNDS OF SALT.

Your Farmer's Body Needs Protection: Healthcare

by SEVERINE VON TSCHARNER FLEMING
Concept: **Patrick Kiley** *Image:* **Laura Cline**

The young farmers movement is growing, and the circle of caring continues to expand. As we work to build a business around our love of farming and a family alongside our practice, we encounter one scary part of growing up: Realizing how deeply critical our own health is to the viability of the farm. As young farmers with brave muscles and big dreams, we invest our best physical years in finding, setting up and capitalizing a farmstead. As entrepreneurs, we take tremendous risks and reinvest the earnings in service to a new small business.

As citizens, we commit ourselves to place and to the performance of an ancient and sacred duty: providing sustenance to our community. But when the operation of all these interlocking systems relies for its longevity on the physical strength and resilience of an individual body, the body of the young farmer turns out to be one of the weakest links in the new food system.

We need healthcare. Many of us cannot afford it. Farming is physical labor with physical risks and with great demands on performance over time. As a nation served by many workers, some unionized, some wearing uniforms, we recognize the importance of retaining skilled practitioners with benefits. Our firefighters, coast guards and electricians are all provided with benefits, and healthcare. Why not farmers? Our enlisted soldiers and their families are provided with

Herniated disc.

Treatment: rest, lumbar exercises, plentiful apprentice help.

Killer Migraines.

Treatment: Massage, herbs, biofeedback, soft voices.

Carpal Tunnel Syndrome.

Treatment: rest, gentle exercises. Sharing in shoveling duty. Easing grip on reins or wheel.

Chronic joint pain.

Treatment: Ergonomic improvements to milking parlor.

Farmer's Lung (hypersensitivity pneumonitis).

Treatment: Breathing exercises, fresh air, no moldy feed room.

Lyme Disease.

Prevention: Daily Tick checks!

Treatment: Antibiotics. Garlic. Licorice.

Seasonal Affective Disorder.

Treatment: Light therapy, brisk morning walks.

coverage for their service. Why not our farmers?

The reclaiming of our local economy will hopefully, in the next decade, be characterized by greater institutional regionalism. This means schools and hospitals buying food from local farms, this means deep partnerships of commerce within residential districts and within agricultural districts. In order to succeed at this level of engagement, the farmers will negotiate the hurdles of liability, red tape and logistics of rescaling. We'll be operating forklifts and mid-sized delivery vans; we'll be scaling up production. We will spend a lot of time resizing, retrofitting and rethinking systems of food production and distribution, in real time, and at real physical risk to ourselves. This is important work. We cannot lose the hardworking members of the team to illness and injury. We cannot lose any fingers or toes. We cannot afford for our farmers to be distracted by financial worry associated with the birth of a child or the infection of a blister. We need to provide health coverage for farmers, young and old, owners and workers, for the longevity of the sector and of the nation.

Lobbying for these issues is crucial. Are you interested in joining our National Young Farmers Coalition and working with partners to figure out possible solutions to the affordable health care situation? Please join the Greenhorns mailing list so that we can keep you in the loop.

HEALTHINESS OF FARMING.

Agriculture, from the constant and regular employment it gives, is most conducive to health. The labor it requires is calculated to knit the frame, and give strength and vigor to the muscular system; besides, it is mostly carried on in the open air, which tends to promote a free and generous perspiration, and give tone and elasticity to the body; while many kinds of mechanical and other employments are rendered unhealthy by the constrained position in which they must be performed, and the close confinement which is necessary in carrying them on. It is obvious that mind and body must be exercised together, in order to promote a regular and healthy growth in both. Man is naturally indolent, and loves his ease; and, were it not for the strong hand of necessity, he would often become idle, useless, and wicked. The constant and regular employment, therefore, which a farm requires, makes it promotive of good morals; whereas, in other employments, where those engaged are more dependent on others, their labor is more irregular. This often leads them to spend a portion of their time in idleness, and thus bad habits are contracted, acquaintance is formed with persons of impure morals, and their characters and standing in society are destroyed. This is one among many reasons why persons who live in cities are more liable to become profane, intemperate, and vicious than those who reside in the country. The business of farming is one which renders mankind more or less independent of each other, but impresses their minds deeply with a sense of their dependence on the Divine Providence; for they know that they may cultivate his land in the best possible manner, and plant it with the best of seed; yet, unless God gives a favorable season, they can raise no crop. The constantly realize the truth of the holy oracle, that "Paul may plant, and Apollos water, but God giveth the increase." There is no employment that leads a man to reflect so much upon the character and works of God as farming, and, consequently, none is so conducive to morality and the practice of virtuous principles.

WINTER ACTIVITY: FORCING BULBS!

"Foxglove"

GW

> *Chapped Hands.*
>
> The following is the best remedy with which we are acquainted : Wash your hands with castile soap ; apply it with a flannel, and if necessary use a brush, in order to get the dirt from under and around the nails and fingers, till they are perfectly clean. The water in winter, if convenient, may be warmed ; then rinse them in a little clean water, and while they are wet rub them well all over with about half a tea spoonful of good honey; then dry them well with a clean towel. This should be done once or twice a day, and always before going to bed.

Activity: Sedentary.
Just sitting, reading, watching TV, listening to radio, playing cards.

Activity: Light
Standing up and cooking, washing dishes, ironing, walking slowly.

Activity: Moderate.
Walking moderately fast, playing ping pong, mopping, making beds, light gardening.

Activity: Vigorous.
Washing and waxing the car, walking fast, bowling, golfing.

Activity: Strenuous.
Running, swimming, tennis, dancing, football, skiing.

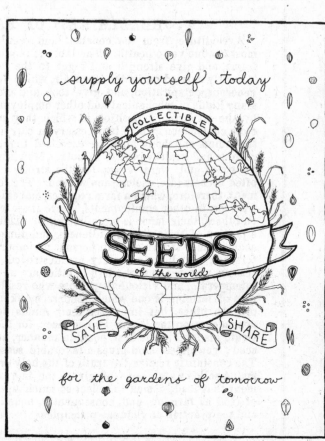

supply yourself today

COLLECTIBLE

SEEDS
of the world

SAVE   SHARE

for the gardens of tomorrow

SG

IN WINTER, EAT CITRUS

EDITORIAL NOTE. Remember, many organic citrus growers, particularly family run orchards are glad to ship a case or a pallet of fruit to you in the month of January and February. Vitamin C can do wonders for your outlook and constitution. In our town, the Elks Club organizes a shipment from Texas of Ruby Red Grapefruit on the bulletin boards of the town. You can call a phone number and order a case. In all likelihood, once your friends learn about your citrus, and taste it, they'll want to order a case from you too. A good excuse for a lemonade party in January. These also make great Christmas presents. You can find the orchard via word of mouth, or through Local Harvest.org, or via the listings of farms from Florida, Arizona, Texas, California local farming organizations or certifiers. It is not rocket science. It is citrus in winter. - SvTF

Vitamin C for the Family

By Mrs. Edith Jenkins Tilton
Home Department Editor

(Mrs. Tilton can be reached by addressing Poultry Tribune. She always welcomes mail from Poultry Tribune subscribers. Being a practical poultry raiser and having taken special poultry courses, she is particularly qualified to give help. Feel free to write to her. Please enclose a stamped return envelope.—Editor.)

VITAMINS, announced for the first time in 1912, were thought to be only one in number, vitamin A, but today we know definitely of six with more in sight. Our poultry raisers have become aware of the great value of vitamin D in their poultry operations.

Water soluble vitamin C is another which we should know, since its presence is absolutely imperative to health and because it is not frequently found in other than fresh foods. Vitamin C is found in fresh

Mrs. E. J. Tilton

fruits, leafy vegetables, tubers and roots and to a slight extent in milk and meats. When the diet lacks vitamin C, scurvy soon develops. Symptoms of scurvy are soreness of the joints, soreness of the gums, loss of weight. Prolonged and aggravated cases which are not remedied by proper diet steadily grow worse and death may result. Scurvy attacks infants fed pasteurized or boiled milk, unless fresh fruit juices, such as tomato or orange, is also given. Adults who live exclusively on canned, dried or cured foods, also become afflicted with scurvy.

Lemons, Oranges, Tomatoes Are Sources

Lemons are known to be very rich in vitamin C. Oranges, also, are another source of vitamin C which both cures and prevents scurvy, but heat and canning destroys the vitamin content of these fruits and often times the fresh foods are beyond the purse of many suffering from the lack of vitamin C. The humble tomato which grows so luxuriantly in our gardens now is rich in vitamin C, retains its vitamin content after canning and proves an inexpensive source of insurance against scurvy.

"Field ripened tomatoes," according to Wisconsin Home Economics research department, "that are canned cold pack method, provided they are used within nine months after canning, are as potent in vitamin C as the same kind of tomatoes fed raw and fresh from the field. Canning open kettle method reduces their vitamin potency one-third. Any kind of tomatoes used a year or more after canning will show marked loss of vitamin C content. To insure the potency of vitamin C of canned tomatoes, it is desirable that they be used inside of nine months after canning. Allowing green tomatoes to ripen at 70 degrees F after picking seems to permit the development of vitamin C to as great an extent as when vine ripened in the field." Left attached to the vine, in storage, the tomatoes will ripen with much more natural flavor.

Busy farm mothers are now insuring their families an adequate and inexpensive supply of the health preserving vitamin C via the route of canned tomatoes. Canned tomatoes lend themselves to many varieties of food preparation. Soups, scallops, meat dishes, salads, gelatin salads and many attractive relishes, conserves and preserves may be made of the fresh tomatoes. Tomato cocktail, strained tomato juice, is a popular and low priced substitute for orange juice for breakfast for both adults and children.

To FARMERS—A Hint.---The writer of these remarks was once a farmer's boy, and speaks from experience when he recommends all farmers' sons to keep a daily register of every thing interesting coming under their observation, relative to their business. The time of planting or sowing crops, with the results of late or early planting appended; the effects of any peculiar mode of manuring; the benefit or detriment from thick or thin sowing; the kind of seed; the time and manner of harvesting; the results of draining, of deep or shallow ploughing, and of numerous other matters, and especially including the cost and profits of each crop, if accurately recorded, would not fail to yield a great deal of interest as well as usefulness. The time of the appearance of birds, insects, the flowering and fruiting of trees, or anything else in relation to nature and its productions, would assist very much the acquirement of knowledge on these subjects, if made a matter of record. I am sure it would be a delightful employment, both at the time, and by its examination afterwards.

Now, all that is necessary is to get a small blank book, with a flexible leather cover, which may be had for a little at any book or stationary store, and rule each page into two columns, the first for the record of planting, sowing, and all other operations during their earlier stages; and the second column for the registry of the results, directly opposite, on the same page. By comparing these results with the operations which produced them, a great deal of valuable practical knowledge would soon be obtained.

Another advantage might result from this practice. When any operation was deferred till too late, and loss was occasioned thereby, make a memorandum of this fact at the proper place in the second column, by the examination of which, the second year, this difficulty might be avoided. Many failures occur from a want of seasonable attention; such a journal would therefore leave an excellent memorandum book to refer to daily the second year, or any other year afterwards, to remind one of what must be done at the time.

Would not this be worth a thousand times its cost, by way of making accurate, intelligent, practical and successful farmers, of lads and young men in the country, besides improving their knowledge of writing?

On Journal Keeping as a Good Farming Practice

by SARAJANE SNYDER, AUGUST 2012

. . . .

As farmers, we work within natural cycles. As modern-day humans, our brains and bodies have lost much of their instinct for operating within these cycles and also the natural skills of observation and memory. Keeping an ongoing journal of any kind can help restore some of these lost neural pathways.

Journal keeping can happen on a wide spectrum from accounting ledgers to artist sketchbooks, but it really only works when it is done consistently over time. Journal keeping is a record that, with continuity and detail, can bring a past farm season or vacation or series of business transactions back to life for a moment. Also, working at a journal with diligent effort over time trains our senses and our memories towards attentiveness and it trains our habits with daily practice.

It can be quite challenging to keep a daily journal. "I'm too busy!" or "Nothing happened today that's worth writing down" or "I can't possibly get down ALL of what happened today" or (gasp!) "I'll do it tomorrow!" Thoughts like these can sabotage the daily journal practice. Meet these thoughts and overcome them! Nothing breaks the flow of a journal like a day without writing (which all too often turns into a few days without writing). A day without a line of writing becomes a void. Even one sentence, the more concrete the better, will usually later serve as a catalyst causing other memories from that day to surface. It also adds something to the days on either side to have at least one thing written down instead of nothing.

At times I have kept many different journals at once, which can either be overwhelming or kind of co-inspiring. I kept a journal for books I read. One for recipes. One for herbs & wild plant uses and identifications. A dream journal. An expense/budget journal. A vehicle log. I've known athletes who keep journals on their training and performance. People keep gratitude journals where they practice giving thanks every day. Some folks trying to lose weight or quit drinking write down every thing they eat or drink.

I kept a farm journal (for which I used one of those fancy day planners that had a week spread out over two pages and ample space to write) where I could keep track of people's days off, meetings, scheduled farm work, the weather, and then any notable events or thoughts of the day. I found it helpful in this case to have the kind of journal where the days are already delineated and if I missed a day it would be an empty hole. That farm journal now lives with the other journals of past farm managers in the office of the farm for future farm managers to look at. Over the years of recording, patterns build. Old ways that have been lost are rediscovered. Keeping a farm journal is definitely a gift to the land.

I also have kept a personal journal for many many years now, and I can say that each time I am in the process of creating the journal it feels not quite right or somehow lacking, but with each passing year, those pages are imbued with more and more meaning and I am grateful to my past self for thinking to write such a thing down. A few years ago I inherited my pop-pop's collection of journals that he kept over the course of 60 years. No, they weren't full of personal reflection and insight. Mostly they recorded the weather and money spent. One of the earliest of his journals, though, was a "five-year journal," which was so compelling I went right out and found one of my own. It's usually a small book, a little bigger than a checkbook, but very fat. Each day of the year has a page and each page is divided into five spaces—one for each
of the next five years. So on year one, you'd go through and give a brief statement on the top space every day for a year. Year two you'd do the same thing, but filling in the second space. By the time you're at year five, you're filling out the bottom space and you can look back on any given day and see what you were doing on that day for the last four years. This is an ideal journal for someone who is starting out on a new part of their life and also, possibly, for someone who doesn't want to write too much.

In terms of content, the most important message is to just be yourself. Another side-benefit of journal-keeping is the slow process of validating your perspective on the world. Honesty and candor will resonate out over the years. Style, wit, good drawing skills, ability to capture the whole story are all skills that will develop with practice. If you are keeping a journal that is to be read by others (such as a farm journal), try to be legible. It's also a gift to your future self to be able to read what you've written. Date everything, including the year. I often try and begin each entry with the date, where I'm writing, and what the weather is like. I also put my name and email address in the front of the book, in case I leave it somewhere.

Yes, you can use a computer to keep a journal too. There are probably even apps to help you along. Unless you are a truly savvy & plugged-in person, I advocate plain old pen-and-paper. It's ok. You can call me old fashioned. Like with anything in farming, experiment with tools and find the ones that work best for your style.
Happy observing!

Farmers And Wildlife

by MEGAN PRELINGER

Farmers are partners in the land with wild animals. The same diverse farm ecology that benefits food production also benefits the entire web of life. Farmers who encourage crop diversity, wind breaks, and patches of wild space on their land are rewarded with a land rich in fellow animal species. Many animals are beneficial to farming, helping to control pests and weeds, pollinating plants, and enriching the soil

The information offered here is aimed at situations in which a farmer encounters an animal in distress. Everyday life in the wild is dangerous for animals, and people are not expected to change that. However, many threats to animals are directly or indirectly caused by people.

Animals can fly into power lines, become entangled in wire or equipment, sustain gunshot wounds, become entangled in fishing line and tackle, and become poisoned by non-food substances. They can also be hit by cars, boats, kites, and tractors, or mauled by dogs, cats, or uninformed people. Animals sometimes also nest right in the middle of a field that needs to be plowed.

Wildlife rehabilitators have developed their healing arts partly in response to the ethical problems posed by the impact of people on wildlife. As a farmer, you can be ready to deal with these situations by knowing who the wildlife care professionals are in your area and being able to reach them.

Here is how to educate yourself: find out who the wildlife care professionals are in your area. There may be a wildlife clinic listed online near you, or your local Animal Care and Control agency may be able to tell you who to contact. A local veterinarian or U.S. Fish and Wildlife officer, or even the police or sheriff may also be able to tell you who to go to. Some licensed wildlife rehabilitators work out of their homes. Have the phone number of your local wildlife rehabilitator(s) on your emergency telephone list.

To be even more prepared, call the local wildlife rehabilitator when you have a quiet moment, and find out what their specialties are. Some wildlife centers treat all animals, others specialize in birds or in mammals.

Read the Guidelines for Safe Handling of Wildlife so you are prepared for emergencies.

It is important to be aware that many laws and regulations govern the handling and capture of wildlife, and that these vary state by state. Look up the phone numbers of the state and federal wildlife agencies in your area, and put those numbers on your phone list as well.

For example, if a flock of White-Faced Ibises has nested in your field, and you need to plow it, there are laws to protect wildlife that govern what you can do about it. Your local U.S. Fish and Wildlife officer can advise you, and can possibly assist you in getting a permit to relocate the nests. If your farm is a haven for endangered species, educate yourself about those species and about any special considerations that may apply.

FOR FURTHER INFORMATION:
THE NATIONAL WILDLIFE REHABILITATORS ASSOCIATION
WWW.NWRAWILDLIFE.ORG
U.S. FISH AND WILDLIFE SERVICE HOME PAGE
WWW.FWS.GOV

For a wasp sting.—Bind on the place a thick plaster of common salt just moistened; it will soon extract the venom. In case of swallowing a wasp, which is a most dangerous accident, it should be instantly attempted to get down a spoonful or more of salt with just water enough to make it liquid. This is a remedy always at hand. Salt and oil would be very useful in such a case, or salt, oil, honey, and vinegar, but there is not a moment to be lost in fetching or mixing what may not be close at hand.

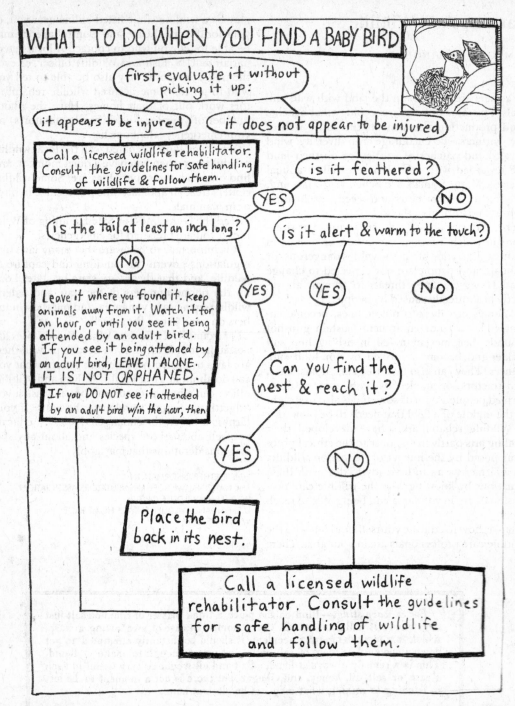

WHAT TO DO WHEN YOU FIND A BABY BIRD

first, evaluate it without picking it up:

it appears to be injured

it does not appear to be injured

Call a licensed wildlife rehabilitator. Consult the guidelines for safe handling of wildlife & follow them.

is it feathered?

YES

NO

is the tail at least an inch long?

NO

is it alert & warm to the touch?

Leave it where you found it. keep animals away from it. Watch it for an hour, or until you see it being attended by an adult bird.
If you see it being attended by an adult bird, LEAVE IT ALONE. IT IS NOT ORPHANED.

If you DO NOT see it attended by an adult bird w/in the hour, then

YES

YES

NO

Can you find the nest & reach it?

YES

NO

Place the bird back in its nest.

Call a licensed wildlife rehabilitator. Consult the guidelines for safe handling of wildlife and follow them.

ARA

Guidelines For Safe Handling Of Wildlife

by MEGAN PRELINGER

1. Human safety is always the most important consideration when handling wildlife.

2. Do not offer food or water to injured or orphaned animals.

3. Do not attempt to handle:
- large birds of prey
- rabies vector species
- predator mammals
- snakes

Without the assistance or supervision of an experienced wildlife care professional.

4. Have ready a box or pet carrier that is:
- dark (draped with sheet or towel)
- well-ventilated
- neither too warm nor too cool
- safe from domestic animals and curious children

5. Have on hand spare sheets, towels, and washcloths, and leather gloves for handling all animals other than small birds.

6. Gently place the injured animal in a dark, well-ventilated box or pet carrier.

7. Wildness can be disrupted by human contact. Wildness needs to be protected. At all times, minimize human contact with wild animals. This includes visual contact. Keep carriers or containers draped with loose sheeting to protect the animals from the stress of seeing people. Do not talk to wild animals, or handle them any more than the situation demands.

8. Transfer your injured or orphaned wild animal into the care of a wildlife professional as soon as possible.

9. Be open to simplicity: If a small bird flies in to your house, you can just shoo it out again. Or pick it up and put it out. Brief, gentle handling will not harm it.

POSTER BY THE BEEHIVE COLLECTIVE
WWW.ETSY.COM/SHOP/GREENHORNS

Faisans.

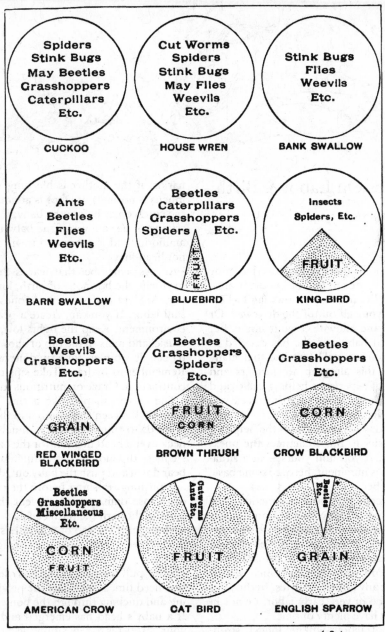

FIG. 107.—Food of some common birds.   (181)

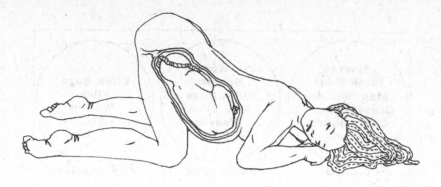

BB

Some Basics On Labor & Birth

by LUCY CHAPIN

What do you do if someone you love has gone into labor and you are 100 miles from the nearest hospital. Or you live in a rural, snowy land where you cannot dig yourself out of the driveway. Or you have called the midwife to your home, but this is your third baby and she has decided to make a precipitous arrival? I realize that many of you reading this almanac are farmers and may have years of experience helping calves and lambs into the world. Trust your instincts; you will know just what to do!

Women should always labor with the help of a midwife or physician. But sometimes the unexpected happens. If you find yourself in a situation where birth is imminent, here are some basic things to remember:

If a mother is feeling sporadic contractions, she may be exhausted or dehydrated. Don't panic. Try to hydrate and get the mother to rest before anything else. If you are in a well-stocked kitchen, try getting the mother to drink a small bit of juice mixed with water, and pinch of salt. This will replace any lost electrolytes. In this situation, she may be in very early labor, or not in labor at all. You have plenty of time.

True labor contractions will get longer, stronger, and closer together. If labor has started, the first thing to do is call for help to determine if she is in labor or not and in what stage. Call 9-1-1 and get to nearest hospital with any of these scenarios: if the mother is bleeding (blood-tinged mucus is normal); if labor is abnormally long; if the cord comes before the baby; if the mother is seizing or has a fever; if the baby is stuck; if the amniotic fluid is green or brown; if the baby is not head-first.

- Always remember that fear is the antidote to oxytocin, the hormone of birth, love, and bonding. Anxiety and stress hormones will actually stall labor. If you can, create a peaceful, homey environment. Keep the lights low, put on some music, and relax in the tub or shower. Warm water can do wonders for a tired uterus.

Remember to hydrate. Take sips between every contraction. Graze on nutritious food.

- The cervix—shaped like a tiny donut—is the gateway between the womb and the birth canal. It needs to thin and stretch, much like a turtleneck over a head in order for the baby to come.

- Unless the mother has an involuntary urge to bear down and push the baby out, do not encourage pushing. Pushing before the cervix is ready can cause damage and prevent the baby from passing through the birth canal.

- Never push the top of a mother's womb to get the baby out. This can cause damage to the uterus.

- Never pull the baby if the baby feels stuck. Babies need time to navigate the pelvis—corkscrewing and ducking through the bony prominences. If a baby's head has emerged and seems stuck, your best bet is to reposition the mother—pulling knees back towards the chest, squatting, getting on all fours. Usually, that is all the baby needs.

- Hand-washing is paramount for the health of the baby and mother. Avoid placing fingers inside the vagina.
- If the umbilical cord comes out before the baby, have the mother get on hands and knees, then place her head and chest on the floor so that her bottom is the highest point. This will relieve some of the pressure off the cord and keep oxygen flowing to the baby. Go immediately to the nearest hospital that delivers babies.
- When the baby is born, wipe any mucus off its face and dry with a warm blanket. If the baby is born quickly, it is possible it will be born in the caul, meaning it is still enclosed within the amniotic sac. You may need to lift the "veil" (amniotic membrane) off the baby's face. It many cultures, being born in the caul is considered auspicious.
- After the birth, dry the baby and place the baby's bare skin on the mother's bare chest. Cover both mom and baby with a warm blanket. This is called skin-to-skin. It is the best way for a baby to thermoregulate and encourages maternal-infant bonding.
- Some babies need their backs rubbed to stimulate their breathing.
- Keep an extra warm blanket nearby. The mother may be "shaky" after birth. It is a normal hormonal response.
- A few minutes to a half hour (usually about 5-10 minutes) after the baby is born, the placenta will need to come out. The mother will usually feel cramping and an urge to push out the placenta. In order to prevent postpartum hemorrhage, the mother must birth her placenta.
- Do not tug on the cord. If there is tension, it means the placenta is not ready to come out.
The best way to ease the placenta out is to have the mother squat. Her abdominal muscles will do the work. She can also use her pushing muscles to help it along.
- If the mother is soaking more than one pad per hour, looking pale, or feeling dizzy, she may have excessive bleeding. The best way to control bleeding is to rub the mother's belly vigorously and encourage breastfeeding.
You can tie dental floss tightly around the umbilical cord in two places and cut between the two strings if you wish to cut the cord. To cut, use sterile scissors or a razor blade.
- Wait at least a few minutes after the birth before you cut the cord so the baby can get its remaining blood supply and nutrients from the mother. It is also safe to leave the placenta attached to the cord and wrapped with the baby until you reach professional help.

Lucy Chapin is a certified birth doula and registered nurse, currently in her final year of midwifery training.

SOUR MILK DOUGHNUTS

No native Vermonters goes without doughnuts for breakfast, even if mother has to get up before-hand to make them. The recipe: 2 cups of flour, ¾ teaspoon soda, 1 teaspoon cream of tartar, ¾ teaspoon of salt, a dash of nutmeg sifted together. Add a beaten egg, ½ cup sugar, tablespoon melted butter, and ½ cup sour-milk (or buttermilk). Knead on the board adding flour, if necessary; cut and fry in deep fat. For Sugared Doughnuts, when cool, shake in a bag with some confectioners sugar.

FROM THE FOOD OF A YOUNGER LAND, FROM THE LOST WPA FILES COMPILED BY MARK KURLANSKY

EDITORIAL NOTE: A New England Farmers Union bumper sticker says it best: "Local food is primary care!" Caring for the particular strains of farmers bodies has become the mission of many alternative healer peoples. In many parts of the country community accupuncture networks are self organizing (www.pocacoop.com). This makes accupuncture more affordable for farmers and others. Stretching, theraputic massage, and strengthening core muscles. - SvTF

Parade of winter rot
Having up-hilled its snowy dent
Tumbles to the daffodill's side

–Mariette Lamson

MARCH

———

CLOUDY

MARCH 2013

zodiac signs

1 — Womens History Month
in 1917 the first federal land bank was chartered
♍ F
≈ ☉ 6:30 / 17:48

2 — ♍⚊ SA
≈ ☉ 6:28 / 17:49

3 — National Unity Day in Sudan
♏ SU
≈ ☉ 6:27 / 17:50

4 — ♏ M
≈ ☉ 6:25 / 17:51

5 — The moon is perigee (closest to the Earth in its orbit) today. Seeds sown at this time tend to be more prone to attack from pests and fungi, especially in a wet climate or season.
PG ♏ ♐ TU
≈ ☉ 6:23 / 17:52

6 — ♐ W
≈ ☉ 6:22 / 17:53

7 — ♐ ♑ TH
≈ ☉ 6:20 / 17:54

8 — ♑ F
♓ ☉ 6:19 / 17:55

9 — ♑ ≈ SA
♓ ☉ 6:17 / 17:56

10 — Daylight Savings Time
≈ SU
♓ ☉ 7:15 / 18:58

11 — NEW
≈ ♓ M
♓ ☉ 7:14 / 18:59

12 — ♓ TU
♓ ☉ 7:12 / 19:00

13 — ♓ W
♓ ☉ 7:11 / 19:01

14 — ♓ ♈ TH
♓ ☉ 7:09 / 19:02

15 — ♈ F
♓ ☉ 7:07 / 19:03

16 — ♈ SA
♓ ☉ 7:06 / 19:04

17 — ♉ SU
♓ ☉ 7:04 / 19:05

18 — AG ♉ M
♓ ☉ 7:02 / 19:06

19 — ♉ ♊ TU
♓ ☉ 7:01 / 19:07

20 — Vernal Equinox. A day of equilibrium, neither harsh winter or the merciless of summer. The Sun will shine directly on the equator and there will be nearly equal amounts of day and night throughout the world.
♊ W
♓ ☉ 6:59 / 19:08

21 — Human Rights Day
in 1960 police killed 69 people at a protest in Sharpville. This holiday was created to ensure that the people of South Africa are aware of their rights and to prevent abuse from happening.
♊ TH
♓ ☉ 6:57 / 19:09

22 — ♊ ♋ F
♓ ☉ 6:56 / 19:11

23 — ♋ ♌ SA
♓ ☉ 6:54 / 19:12

24 — ♌ SU
♓ ☉ 6:52 / 19:13

25 — ♌ M
♓ ☉ 6:51 / 19:14

26 — ♌ ♍ TU
♓ ☉ 6:49 / 19:15

27 — FULL
♍ W
♓ ☉ 6:47 / 19:16

28 — Today, Mercury and Saturn are in trine. In this position, the planets are in constellations corresponding to the same plant part. If this part is different than that designated by the position of the moon at that time, the trine can override the moon's influence.
♍ TH
♓ ☉ 6:46 / 19:17

29 — ♍ ⚊ F
♓ ☉ 6:44 / 19:18

30 — ⚊ SA
♓ ☉ 6:43 / 19:19

31 — PG ♏ SU
♓ ☉ 6:41 / 19:20

THE NEW OLD TRADITIONS
THROUGHOUT THE YEAR

SPRING EQUINOX: "The Birth of the Sun"
(On/around March 21st)

The Spring Equinox marks the passing of the
torch, the time when the rhythms of the planet
are guided more and more by the Sun, for it is
on this day that the Sun is born. (ALL: Hooray!) From now until the Midsummer night
the Sun will become the most prominent influence in our daily lives and will inform how to
spend the next few months. The Spring Equinox is a time to celebrate birth, and is thus a
time to honor what lives. So too is the Spring
Equinox a time to celebrate equality, as it is on
this date that the lengths of both day and night
are equal. It is a time of blending and merging,
of alchemical unions, and playful couplings.
The Spring Equinox is when two become one.

MOCK HOROSCOPE

(From Poor Robin's Almanack for 1690)

Aerial Mississippi

by FRANCESCA CAPONE

((
))
((
))
((
))
((
))
((
))
((
))
((
))

Air Rises

by ROSMARIE WALDROP

Air rises
blue
irresistible with distance
place to
stay
immobile
a long time
at the edge of

*Ceres, goddess of agriculture,
grain crops, fertility and
motherly relationships*

KWR

Cutting of Grain.—It is the opinion of experienced husbandmen, that wheat should be cut down some days before it is fully or dead ripe. The grain hardens well in the sheaf; the sample is better; and there is nothing lost in measure by this mode of management. The harvest thus begins earlier, and its labours are more equally distributed. Barley ought likewise to be cut down before it is ripe; otherwise the straw becomes brittle, and much loss is occasioned by the heads breaking off. Though oats are reckoned a hardy grain, yet the more early varieties, being liable to damage from high winds, or from exposure to much heat, ought to be cut as soon as they are nearly ripe, in order to lessen the risks to which they are exposed. An experienced farmer states, that all sorts of grain ought to be cut whenever the straw, immediately below the ear is so dry, that, on twisting it, no juice can be expressed; for then the grain cannot improve, as the circulation of the juices to the ear is stopped. It matters not, that the stock below is green. Every hour that the grain stands uncut, after passing this stage, is attended with loss.

The Corn Mother

by KATIE RAISSAIAN

The corn mother wandered the fields and meadows dressed in green. Wherever her foot fell, crops sprang up. November seas roared.

The moon-bull watered the seed-corn with its tears. The sow-goddess made the corn grow tall. The corn mother ruled over seasons and the fields. When she was happy, crops never failed. When anguished, the earth turned cold and hard and hunger spread.

From the corn mother's worship of the moon-bull came the cowherd and the spring. From her copulation in a thrice-ploughed field came the god of plenty.

The cowherd, keeper of the pickax, plough, and yoke, showed his people how to till the land. When he died, his body was severed into twenty parts. His limbs were carried to the corners of the earth. Corn fields flowed from his fallen flesh.

When springtime fell beneath the earth, the corn mother left the great white mountain in search of her. She scoured the earth for nine whole days. She cast a babe into a sacred fire. She transformed into mare and suckling pig. She wept within a temple. But still nothing grew for one whole year. The rimy ground remained.

Within a grove of coal-black poplars, the lone white aspen wept. Inside her, seven fruit seeds swelled. From across the blackened waters, she heard the corn mother's call. The spring beseeched the death-god to be freed. He bargained for her life: For nine months, she would bloom again above; for three, below, she would be his wife.

From atop his cornucopia, the god of plenty watched. His horn sounded to signal spring's return to earth. The flora rejoiced and all was abundant. Mock-mothers laid in thrice-ploughed fields, performed mock-births in the corn-babe's honor. Their mimicking the corn-mother ensured a healthy yield.

At midnight on the last day of harvest, the corn spirit haunts the fields. She is said to be trapped in the last sheaf of corn and is flogged by pickers 'till the final grain falls. She is drowned in water so that rainfall comes. She is lain in a barn to keep mice at bay. She rests in a basket for over a year. She is hung in a barn until threshing is done. The last thresher is flogged in the folds of her skirt.
She is a pig, a horse, a cow, a cat,
a hare, a bird, a wolf, a dog,
a goat, a bull, an ox, a fox,
a cock, a goose, a quail.
She is rattle, fan, maer, maiden, cailleach, crown, and lantern. She is the wind, the bell, the colors blue and red. She makes coyotes of jealous maidens.

Look to the soil to know the mystic origins of the corn mother. Lament the doomed corn-spirit; run her from the fields. Seduce the happy spirit with ox-blood and peony. Crop her limbs with a golden scythe and bind a snake around her waist. Entreat her from the meadows when the last ear falls, and the cold, hard cover of winter comes.

JB

The Landscape Of Corn
AND WHEN NOT A CORNFIELD
IS A CORNFIELD

REPRINTED WITH PERMISSION
FROM THE CENTER FOR LAND USE
INTERPRETATION
NEWSLETTER WINTER 2006

The corn is gone from downtown Los Angeles. The harvest was the last event in a nine month project known as Not A Cornfield, which involved planting 32 acres of corn in a former railyard brownfield near Chinatown known generally over the years, for some reason, as the Cornfields. Not A Cornfield was conceived by the artist Lauren Bon, as an artwork, or, more accurately, as a nexus for a network of converging activities, events, lectures, screenings, and artforms.

The land, between the LA River and Downtown, is scheduled to become a state park. In the meantime, for less than a year, Not A Cornfield LLC took over the space, starting in the summer of 2005. The project brought in hundreds of truckloads of dirt to lay on top of the brownfield ground, then formed furrows, lined with irrigation pipe, and planted corn.

While the corn grew from July to November, Not A Cornfield became a social space, free and open to the public, with scheduled and unscheduled activities. Films were screened, talks and discussions were presented, music was performed, and people caroused, in clearings in the corn, and at the construction trailer (and yurt) compound at the entrance to the site. A central path, cut through the middle of the oblong field, enabled visitors to walk through a corridor of corn for nearly half a mile, towards the silhouetted downtown skyline.

As the corn aged, dried, and turned brown, it was like a Halloween maize maze in January. Openings and new paths were cut into the fields, some resembling crop circles. Some stalks were grouped into bushels, or shocks, to help define large, circular galleries for events, and elaborate lighting was installed that mirrored constellations during the winter solstice. The corn was picked, and an estimated 137,694 ears of corn were hung on the fence along the commuter rail tracks.

When it was time to clear the field, a John Deere 9660 combine came in and mowed the field down, churning up the stalks into mulch that was raked, baled, and used to construct a monument, known unofficially as "corn-henge." The ears of dried corn were also fed into the machine and the decobbed kernels were decanted from the combine's hopper into large bins, then sifted, and bagged for distribution as seed corn to homeless shelters, community gardens, and correctional facilities.

While it is sad to see this unusual and dramatic physical and social artwork gone, it led us to ponder the larger ideas of corn in our lives, and in America. . .

THE AMERICAN LANDSCAPE OF CORN

The Lower 48 states of the USA is 1.9 billion acres in size. Of that, about 30% (600 million acres) is forested, another 30% (580 million acres) is grassland and rangeland used by cattle and such. 10% is called "nonproductive," from an agricultural point of view, places like wetlands and deserts. 20% is cropland (450 million acres). On this, corn is America's largest crop, covering 81.6 million acres, about one quarter of all crops, an amount that adds up to about 4% of the land cover of the Lower 48. This is about the same amount of land that is urbanized/suburbanized. This is also almost half of the world's production of corn.

80% of the corn produced in the USA goes to feed livestock, especially cows. The American livestock industry is the largest global consumer of corn, but 15-20% of our corn is exported to other countries, like Japan, where it feeds Japanese cows.

The biggest corn processor in the world is the Archer Daniels Midland corporation, headquartered in Decatur, Illinois. They have 205 manufacturing facilities worldwide, 13,000 railcars, 1,200 trucks, and over 2,000 river barges (said to be the largest fleet in the world) for moving their product around. Three distinct plants operate at the Decatur complex, connected to each other by pipelines. The East Plant at Decatur is probably the largest corn products factory in the world. It processes corn at a rate of 600,000 bushels a day.

What's a bushel? 4 pecks. And a unit of volume, 1.2 cubic feet, about 35 liters. A semi tractor truck can carry 920 bushels.

Though the fuel additive ethanol has recently surpassed high fructose corn syrup as the largest non animal-feed use of corn, the largest direct consumption of corn by humans in the USA is by ingesting corn syrup. High fructose corn syrup is found in numerous processed food products in the United States, where, unlike in other countries, due to the economy of large scale industrial production by companies like ADM, it is a less expensive sweetener than other sugars, like cane and beet. The vast majority of corn syrup is not eaten, but drunk in the form of soda pop, of which 12 billion gallons are consumed every year in the United States. The average American drinks 15 fluid ounces of soda a day, with about 14 teaspoons of corn syrup in it. A 48 oz "big gulp" has a full cup of corn syrup. It adds up to about 70 pounds of corn syrup per person per year.

The largest supplier of corn syrup sweetened drinks is the Coca-Cola Company of Atlanta, with over 40% of the nation's carbonated soft drink market. Their biggest product, Coke Classic, is still the most consumed sweetened carbonated beverage in America. Coke's other brands include Sprite, Minute Maid and Nestea. Not far behind, with just over 30% of the market, is PepsiCo, headquartered in Purchase, New York, up the river from Manhattan. In addition to the cola rival Pepsi, the company owns Mountain Dew, Slice, the Sobe drink line, Gatorade, and Tropicana (there is a lot of sweetener in orange juice too).

Corn as solid, human food comes mostly in the form of corn chip snacks. About a billion bags of tortilla and tostada snacks are sold each year, as part of a $2 billion corn chip industry. The Frito-Lay brand dominates the industry, with 80% of the market share. Frito-Lay, headquartered in Plano, Texas, was formed by a merger of Elmer Doolin's "Frito" fried corn snack company and Herman Lay's potato chip company, in 1961. Based in Nashville, Lay's company grew by purchasing production plants and distribution networks in the southeast, consolidating what was generally a regional industry of small local producers. By 1956, it was the largest potato chip company in the country, but even so, it had just over 1,000 employees. Lay was the exclusive distributor of the Frito snack, the primary product of Doolin's Frito Company, since 1945. As Lay's distribution network grew, so too did the popularity of Fritos. Following the 1961 merger, the company expanded its network to cover the whole nation.

Frito-Lay owns the three most popular corn chip brands in the country, Fritos, Doritos and Tostitos. They also own Rold Gold pretzels, Lays potato chips, Ruffles, Funyons, and Cheetos. The company operates hundreds of distribution centers and dozens of plants across the country, where they manufacture most of these products together. Frito-Lay is consolidating their production, movong more activity to their largest and most technologically advanced plants, including those at Lynchburg, Virginia; Bakersfield, California; Fayetteville, Tennessee; and Jonesboro, Arkansas; as well as at some older, but high-performing plants in Killingly, Connecticut; and Perry, Georgia. Currently, Frito- Lay's plant at Frankfort, Indiana is the largest "salty snacks plant" in the world.

Notably, since 1965, Frito-Lay has been owned by PepsiCo. Nothing makes you thirsty like a cornchip.

· ·

Persephone, goddess queen of the underworld and of spring growth

KWR

FLATBREAD SOCIETY IS RISING

HANDS KNEADED

BREAD IS LIKE CLAY
DOUGH KNEADS HANDS

AF

Ode To The Sea In Spring Fevers

by CATHERINE HANNIGAN

The moon looks down if waning in the sky
and the sea opens its cavernous gut, the most wild

expanse resonates peace: you are the witness
of something unknown and bigger than the self.
The graveyard of lost intentions, the perspective
of loneliness, and the bohemians stretch out their toes
barely touching the rolling salty foams.
Go deep but not too far.

The clouds move quickly
begging the heart to joyfully collapse
perfecting the mind forever and still.

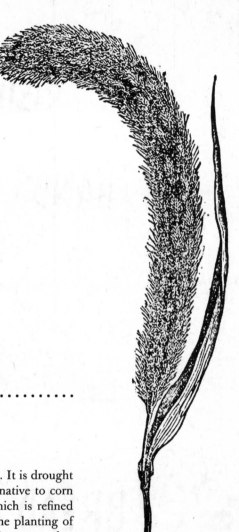

. . . .

EDITORIAL NOTE: This crop is mainly grown in Africa. It is drought tolerant, sometimes grown in the midwest as an alternative to corn and also harvested in the southeast for molasses-- which is refined from the sap in its crushed stems, like sugar cane. The planting of sorghum in a polyculture is common as an "insurance" crop-- so that when there are drought conditions, like in 2012, that there is still food to eat. While the major use of sorghum in the United States is as bird-food, it is very healthy, similar to millet, useable with other flours in bread, and can be made into booze. - SvTF

"Sorghum"

APPLES.

This fruit being too generally considered only as affording a beverage; our farmers are apt to be indifferent to the species which they raise. The natural fruits, it is true, often make the best cider, but it is not uncommon to see a farmer who may make twenty or fifty barrels of cider, unable to pick out a single barrel of fine apples which he can preserve to a time when he wants them most, in the spring of the year, when they are as salutary as they are agreeable. The scions of apples may be procured and sent to any distance in March, and till the tenth of April, and if well taken care of, by being plunged in clay or moist earth, they may be inserted from the twentieth of March to the tenth of June. Any farmer might soon learn the art of engrafting, and their old orchards will furnish stocks. Trees are often preserved and renewed by heading them down and grafting them, if this is done judiciously.

The sorts which we should recommend for general cultivation are,

The *Rhode-Island Greening*, a good *fall* and *early winter* apple.

The *Nonsuch*, a red apple, excellent, and very late keeping apple.

The *Nonpareil*, a Russet apple, early in winter.

The *Newtown Pippin*, a good, hard, late keeping fruit.

The *Spitzenberg*. This is a fine fruit, and will keep sound till May or June.

The *Roxbury Russetting*. This is one of the best known and most valuable fruits. It is not fit to eat till February, and is very easily preserved till June.

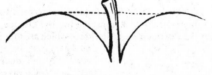

The *Baldwin Apple*, (recently brought into notice, though it has been in the country probably for many years,) is a very valuable fruit, beautiful, and fine flavored, and will keep till the last of March.

There are at least fifty sorts of good apples, beside those above specified; we have only noticed those, which would be most extensively useful as winter fruits. We have selected those which always command a price in market.

BALDWIN

EDITORIAL NOTE: Few gooseberries are grown commercially in the United States, it is not a fruit we expect to find at our markets and grocery stores, but it is a very early tart fruit. Delicious in jams, compotes, and fruit salads. It is hardy, quick growing, prolific, and suitable for inclusion at forest edge, in a hedgerow/shelter belt. There are prickles, yes, but those translucent green sourlings are worth it. - SvTF

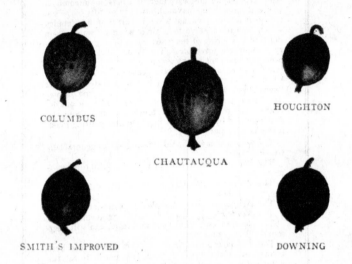

COLUMBUS

CHAUTAUQUA

HOUGHTON

SMITH'S IMPROVED

DOWNING

VARIETIES OF THE QUINCE.

The two outlines here given, show, in a reduced form, the two principal old varieties of the quince—the round representing the Orange or Apple variety, and the other the Pear quince.   The trees which are found throughout the country have in many instances, been obtained from seed, and hence present a considerable variety of form, with less of quality.   Rea's Seedling or Rea's Mammoth is a very large and improved variety of the Orange, being one-third or one-half larger and of excellent quality. The tree is a strong grower, with large foliage. The Angers quince used for pear stocks, bears a fruit similar to the Orange in appearance, but is a little later, and slightly harder in texture. The Fontenay or Paris quince does not bear so good a fruit as the Angers, but more resembling the Pear quince in form and texture.

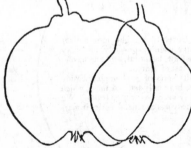

Fig. 24.—*Orange Quince.* Fig. 25.—*Pear Quince.*

THE ILLUSTRATED ANNUAL REGISTER OF RURAL AFFAIRS
AND CULTIVATOR ALMANAC FOR THE YEAR 1869,
THOMAS, J.J. (JOHN JACOB) WWW.ARCHIVE.ORG

To Clean Carrot Seed.—H. Knell, Jo Davies Co., Ill., answers to a question in the April No. of the *American Agriculturist*: "I wish to state that in Germany, we treated carrot seed in the following way: After the seed is gathered, it is put in an airy place to get thoroughly dry. It remains there until time can be spared in winter to pack it in bags; it is then dried in or over a baker's oven; after this it is beaten with a threshing flail for a few minutes, which not only loosens the outer skin, but also the little spines attached to the seed. Then by running through a fanning mill you get cleaner seed than can be procured by any other method."

NEW, FARM-BASED,
REGIONALLY ORIENTED
SEED COMPANIES

GROUNDSWELL SEEDS, ME
SYNERGY SEEDS, OR
ADAPTIVE SEEDS, OR
PEACELING SEEDS, OR

NC

Excerpts from Massachussetts Agriculture, 1868

What is a farmer to do, who has not got it of his own, and cannot or does not dare to borrow it? Well, this is a fair question and deserves an answer equally so. Thou shall have it. He must farm at a great disadvantage all his life, or until he acquires it, as compared with one who has it. Is this hard and discouraging?

If it seems so, recollect if you please, that no one is to blame. It is true, and we are bound to state facts as we find them. Farming in this particular, does not differ essentially from any other business. Here at least, the conditions are the same; and if it seems discouraging to a young man just starting in life, to be told that capital is essential to success, let him bear in mind that it is not so in farming only.

· ·

The truth is, the world is greatly indebted to those who have thought and written upon agriculture. The improvements in stock, fruits, crops, implements and methods of farming are made by men who read and think...in saying that the farmer should read and think upon his calling, we only affirm that in common with every other man who pursues a regular vocation, he should thoroughly understand his business in all its details and relations, and should make the most of it and himself.

Secretary Fling quotes the saying of a celebrated painter, who, on being asked what he mixed his colors with to render them so perfect replied, "I mix them with brains."

That is the material with which every man should mix his colors, and the farmer no less than any other man. He should rear and use his animals, fertilize and cultivate his fields, select and dispose of his crops "with brains" and for those purposes he should use both his own brains and those of other men to as great an extent as possible. Let him, therefore, cooperate in the establishment of farmers libraries and reading clubs. Let him take and read the best agricultural journals...and he may enjoy as much of Eden here in Essex county, Massachusetts, as any of Adams race have enjoyed since the gates of Paradise were closed.

· ·

Another element of success with the farmer is a resolute, determined and persevering industry in one and the same direction. The success of man is only limited by the amount of persevering thought and determined resolution he possess. To the persevering man there is no doubting or waiting, he knows no tomorrow or next week, but it is one determined and resolute now.

Mature your plans, work with a calculation beforehand. Have a fixed object in mind at the beginning of the year, and bend all efforts towards that object... Close planning will save much hard work.

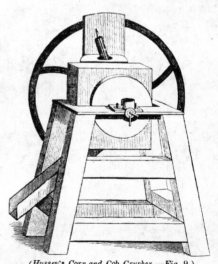

(Hussey's Corn and Cob Crusher.—Fig. 9.)

Every provident parent is anxious to see his children settled in some business of life, that promises to confer wealth and respectability; and every young man, who aims to arrive at future
and honorable distinction, is anxious to select that employment which is most likely to realize
his wishes. It is with a view to enable both parent and son to act wisely in this matter, that we
propose to point out some of the advantages which agriculture holds out to those who embark
in its pursuits.

We propose to consider agricultural employment under the following heads:

1. As a means of obtaining wealth;
2. As a promotive of health, and the useful development of the mind;
3. As a means of individual happiness, the great pursuit of life;
4. As a means of enabling us to fulfill the high objects of our being; of performing the duties
which we owe to our families, our country, and our God.

- Jesse Buel, Farmers Companion (1839)

FROM AMERICAN GEORGICS, YALE UNIVERSITY PRESS

LE

The Handing Down: Reactivating Eco-Gastronomic Heritage

by KATHARINE MILLONZI

"IN THE CIRCLE OF LIFE THE OLD GIVES WAY TO THE NEW AND THE NEW GIVES WAY TO THE OLD." - HOPI/HAVASUPAI /TEWA GRANDMOTHER MONA POLACCA

"THE HEART EATS A PARTICULAR FOOD FROM EV-ERY COMPANION: THE HEART RECEIVES A PAR-TICULAR NOURISHMENT FROM EVERY SINGLE PIECE OF KNOWLEDGE." -RUMI II, 1089

Our food does not come from nothing or nowhere; our food is the outcome of many combined material and social elements. Through the act of eating, we come into daily communion with the earth's gifts. Through the act of eating we all directly collaborate with and impact the Earth's resources that she offers us for our sustenance. Through this understanding of food and the earth, the natural relationship of gastronomy to ecology emerges, and the parameters of an ecological gastronomy are illuminated.

The very small is the very beautiful, wrote the prescient economist E.F. Schumacher. The small interplays of each ecosystem –for example, those that exist between microscopic soil organisms and little seeds – represent the beauty of the barely perceptible, the elemental infrastructure that forms all life. In order to inform future generations of context-appropriate, human scale food systems, we must learn to know again the very small details of our landscapes. We must come again to understand the contextual relation of product to source. At this small scale, nature is the measure and we safeguard the indigenous interconnectedness of human and ecosystem relations.

What we eat, and who we are, is inherited from those who came before us, and what they ate. We are the results of diverse landscapes, certain climates and distinct communities who created unique foodstuffs within those settings and passed on appreciation of particular foods rooted in place. The lineages between humans and their physical environments establish our eco-gastronomic heritage. This type of heritage tells the story of mankind's relationship with food – from seed to animal, soil to mouth, plow to plate.

Our focus on survival in the future distracts from the fact that we are experiencing a crisis of heritage. The older practices that safeguard nourishment at many levels are being lost while we anxiously look ahead. We have disconnected from our physical relationships with the basics of nourishment: air, water, fire, earth and ether. We have also become increasingly disconnected from our relationships with other humans, who not only ensure our food sovereignty but also provide enjoyment and pleasure around meals.

In any country, the preservation of the genetic and materials substances of agricultural heritage - seeds, land, and species - is critical. For example, as our elders teach us which varieties yield well in drought, which meadows the bees like best, which vegetables work for canning and fermentation techniques, we ensure our rightful inheritance to the nourishment this knowledge can provide us in the future. Preparing vital, nutrient dense foods, such as those grown in our own gardens and made by hand in our kitchens, encourages the transfer of past-held knowledge from one generation to the next.

Yet, gastronomic heritage efforts cannot be limited to the conservation of material resources alone. Saving edible species without the corresponding social knowledge of these species is of limited use. Our gastronomic heritage goes beyond material possession to the force of immaterial human knowledge carried through the ages. It is through the building a community of shared concern and purpose around crops that the parameters of community-based food culture are built. Land-based skill sets, such as the use of farm instruments, must be cultivated alongside crops. Our agricultural heritage is based in the labor of communal harvests, in which villages and regions perpetuate unspoken tricks of the trade, and end a season's work with conversation over shared meals that glorify the harvest and perfect recipes. Collective, rural, place-based stories are also part of our heritage, as these stories relay intergenerational understandings of our dynamic co-evolution with the plants and animals upon which we depend. Within a small-scale framework of food production, the common goal of the common good depends on cooperation, association and small talk.

Through revitalizing our food tradition a great opportunity is at hand: the opportunity to engage with our heritage, backwards and forwards. As we begin again to know our neighbors, and return to a community –scaled food culture, we return to the craftsmanship of contextual cuisine, through practices rounded with time and reverence. These practices make up what Wes Jackson calls the "familial and communal handing down of an agrarian common culture". In practice, by offering tips from the past, ecological gastronomy presents possible solutions to humanity's current,

and future, interrelated ecological, economic and social crises. Contextually specific human knowledge of food transfers the skills and spirit necessary to humanity' s future sustainability, because this type of knowledge reaffirms communal investment in sustainable livelihoods.

So, with an eye ahead towards our children's lives, we must nod to those who have come before us. The work ahead is to re-value and re-member the land-based wisdom of our ancestors: indeed, our future food and farm economies depend upon concerted efforts towards a re-activated experience of our agricultural - and therefore gastronomic - heritage. In deepening our relationship to gastronomic heritage, we move towards the re-establishment food cultures grounded in nature's economy. In doing so, we dedicate ourselves to a type of nourishment that extends far beyond our mouths; we reinstate the basis of true intergenerational equity, of true inheritance, and the promise of a truly renewed, truly plump economic life.

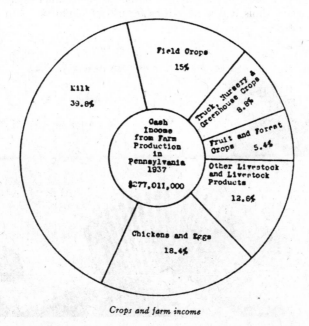

Crops and farm income

Beards

by ELIZABETH HARTSIG

Men have been growing beards lately.
You could call their beards friendly or patriarchal but I would like to think about their general
usefulness, especially in the case of the beards of farmers.
A garlic braid, a hay fork, a welding torch— a man farmer can put so many things
in his beard.
Young horses, scared of shadows, who want tenderness.
A man farmer will even put his own enormous hands into his beard.
All these things can enter a man's beard's deftness. His beard's dominion.
But his deft beard's dearest hope is to hold the cold water that would fill the hollow well.
Men grow so many beards.

Women grow the slenderest, wispiest beards.
Maybe they are like the swift spirits of animals.
Their beards do not pass muster.
Because of this women farmers have sewn pockets all over their clothes.
A pocket is to be kept expansive like any kind of soul.
It can hold a button. A blade. A saved seed. One true, essential thing.
Some of the women's pockets are hidden behind their knit vests.
These pockets nestle to their warm stomach skin.
A woman farmer cannot put sleepy children, or a yarn swift, or a field of light
in her beard.
Slim-bearded, pockets full, boots wearing thin— a farmeress must gather so much in her arms.
For that is what a woman's arms are for. Her arms are also for kindness.
But her old, un-ironed brain— is for thinking.

THE TOMATO AND ITS USES.

EVERY body cultivates the tomato, and every one who has not deliberately made up his mind to be ranked among the nobodies has learned to eat it. There is a great deal of fashion in this, it must be confessed, but it is not often that fashion is active in forwarding so good a work; for if the opinions of numerous M. D.'s of great celebrity, are to be allowed of any weight, there are few things more conducive to health than a liberal use of tomatoes. The fruit has long been extensively used in Italy and the south of France, and within a few years its cultivation as an article of luxury, if not of necessity, has spread over the greater part of Europe and the United States. The fruit is the best in a warm climate, where it has an acidity and briskness unknown in a colder one. In our southern states the fruit is finer and the flavor richer, than in the northern ones; still in these last, abundance of tomatoes of excellent quality are grown, where proper precautions are taken to give them an early start in the spring. Frequent inquiries are made by those who have but lately commenced their cultivation, as to the best modes of cooking or preserving them. To answer these inquiries fully, it might be necessary to consult the taste of each individual, and to gratify these tastes as far as possible, we have collected from various sources, aided by our own experience, such directions for the use of this fruit, as will probably meet the wishes of most.

TOMATO CATSUP.—Use one pint of good salt to one peck of sound ripe tomatoes. Bruise them, and let them stand two days; then strain them dry; and boil the liquor until the scum stops rising, with two ounces of black pepper, the same quantity of spice, one ounce of ginger, one of cloves, and half an ounce of mace. Strain through a sieve, bottle and cork tight.

DAILY USE OF TOMATO.—Slice up as you would cucumbers, good tomatoes after peeling them, with salt, vinegar, and pepper. Some prefer covering the tomatoes when so sliced, with sugar. Either way they will prove palatable and healthy.

TOMATO SOY.—Take ripe tomatoes, and prick them with a fork, lay them in a deep dish, and to each layer add a layer of salt. Let them remain four or five days, then take them out of the salt, and put them in vinegar and water for one night. Drain off the vinegar, and to each peck of tomatoes put half a pint of mustard seed, half an ounce of cloves, and the same quantity of pepper. The tomatoes should be put in a jar, with a layer of sliced onions to each layer of tomatoes, and the spices sprinkled over each layer. In ten days they will be in good eating order.

TOMATO OMELET.—Slice and stew your tomatoes. Then beat up half a dozen new laid eggs, the yolk and white separate; when each are well beaten, mix them with the tomato—put them in a pan and beat them up, and you will have a fine omelet.

TOMATO AS A RELISH FOR BEEF STEAK.—Wash them clean, cut them once in two, lay the inside upon the bars of the gridiron, set them over pretty hot coals for about ten minutes, then turn them over, sprinkle them with salt and pepper, renew the coals, and set them to broil fifteen or twenty minutes longer; when taken up, butter them or eat them with gravy, as best suits the taste or convenience.

STEWED TOMATOES.—Take ripe tomatoes, slice them, put in the pot over the fire without water. Stew them slow, and when done, put in a small lump of butter, and eat as you do apple sauce. If you choose, a little crumb of bread, or fine crackers, may be added.

ANOTHER MODE.—Take the tomatoes and pour boiling water upon them to make the skin come off easy, let them stand three or four minutes, and then peel them. Cut them open and take out the most of the seeds, as too many of these cooked makes the tomatoes too astringent. To one dozen good sized tomatoes, put a small tea spoon full of salt, and a large spoonful of sugar. Stew them from three-quarters of an hour to an hour and a quarter, according as how fast they cook. Stew them down so as to leave very little syrup, and be sure to serve them hot; a shallow vessel is the best to cook them in when stewed in this way.

A Long Time in Coming

Spanish explorers first discovered tomatoes growing in Central America in the 1500's. They brought the seeds back to Europe, and the tomato became very popular—for a while. But then it got a bad reputation. And only in the last 150 years has the tomato been used as food in all parts of the world.

Tom's Tomatoes

In 1781, Thomas Jefferson became the third President of the United States. But 20 years earlier, Jefferson was one of the first persons in our country to grow tomatoes—not for eating, however, but as a plant decoration.

Thank You, Colonel Johnson

The fact that we eat and enjoy tomatoes today is due, in part, to a man named Colonel Robert G. Johnson. For years many people believed that if you ate a tomato, you would die before morning. Colonel Johnson decided to prove they were wrong.

On September 26, 1820, at 12 noon, Colonel Johnson stood on the courthouse steps in Salem, New Jersey. In front of hundreds of people, he ate—not one—but a whole basketful of tomatoes.

Many people thought he had gone crazy—until, that is, he not only lived, but didn't even get sick.

INVASIVE SPECIES OF NORTH AMERICA *by* LUCY ENGELMAN

INVASIVE SPECIES ARE AGGRESSIVE NON-NATIVE SPECIES WHOSE INTRODUCTION CAUSES HARM ECONOMICALLY, ENVIRONMENTALLY &/OR HARM TO HUMAN & ANIMAL HEALTH. INVASIVE WEEDS REDUCE CROP YIELDS, LIMIT PLANT VARIETIES FOR LIVESTOCK FEED, CROWD DESIRABLE VEGETATION, & CONSUME FERTILIZER & WATER.

IDENTIFY

PINK PETALS
(RARELY CREAM COLORED)

BLACK TIPPED BRACATS

SEEDS

INVASIVE PLANT
SPOTTED KNAPWEED

INTRODUCTION: 1890

LOCATION: VIRTUALLY EVERY STATE (ESPECIALLY DISASTEROUS IN IDAHO, MONTANA, WASHINGTON, & MINNESOTA)

EFFECTS: LIVESTOCK WON'T EAT THE PLANT UNLESS THERE IS NOTHING ELSE
-- SLOWS GROWTH OF NATIVE PLANTS
-- CAN PRODUCE THOUSANDS OF SEEDS A YEAR. ONCE A SEED BANK IS ESTABLISHED IT TAKES YEARS TO ERADICATE

INVASIVE FUNGUS
ASIAN SOYBEAN RUST

INTRODUCTION: 2004

LOCATION: PRESISTANT IN FLORIDA, GEORGIA, ALABAMA, ARKANSAS, TEXAS, MISSISSIPPI, LOUISIANA

EFFECTS: FUNGUS CREATES SMALL SEEDS & REDUCES YIELDS. ALMOST ALL COMMERCIAL SOYBEAN GROWERS ARE SUSCEPTIBLE

LIGHT-TANNISH COLORED SPORES
(TYPICALLY ON THE UNDER-SIDE OF SOYBEAN LEAVES)

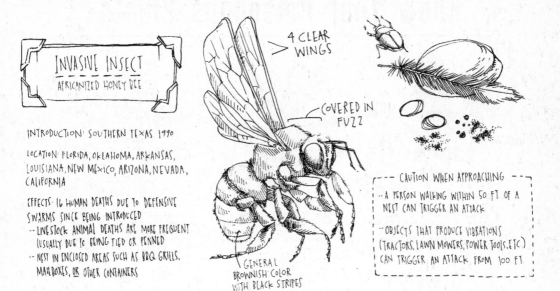

INVASIVE INSECT
AFRICANIZED HONEY BEE

4 CLEAR WINGS

COVERED IN FUZZ

INTRODUCTION: SOUTHERN TEXAS 1990

LOCATION: FLORIDA, OKLAHOMA, ARKANSAS, LOUISIANA, NEW MEXICO, ARIZONA, NEVADA, CALIFORNIA

EFFECTS: 16 HUMAN DEATHS DUE TO DEFENSIVE SWARMS SINCE BEING INTRODUCED
-- LIVESTOCK ANIMAL DEATHS ARE MORE FREQUENT (USUALLY DUE TO BEING TIED OR PENNED)
-- NEST IN ENCLOSED AREAS SUCH AS BBQ GRILLS, MAILBOXES, OR OTHER CONTAINERS

GENERAL BROWNISH COLOR WITH BLACK STRIPES

CAUTION WHEN APPROACHING
- A PERSON WALKING WITHIN 50 FT OF A NEST CAN TRIGGER AN ATTACK
- OBJECTS THAT PRODUCE VIBRATIONS (TRACTORS, LAWN MOWERS, POWER TOOLS, ETC) CAN TRIGGER AN ATTACK FROM 100 FT

EDITORIAL NOTE: The list of invasive species, pernicious agricultural weeds, pests, aquatic cloggers and disease vectors continues to grow as international movement of goods and products continues and ecosystems are degraded. There is only so much we can do address the invasive species present in the places we live and to preserve biodiversity. Meanwhile, genetically engineered mosquitoes are being released experimentally. Our current mon-mega-systems are almost designed to breed superbugs resistant to antibiotics and supersedes resistant to pesticides.

Practically, the first step is knowing which species are considered invasive, and then learning to manage them. Professionals in this field will recommend both chemical and mechanical methods for removing or controlling species. County Extension offices offer training about the plants and animals in your area. Those that most jeopardize economic activities will be the most likely candidates for study and easiest to learn about. For instance, Japanese Knotweed (which has edible shoots and makes great honey) is usually sprayed with Glyphosate after it has gone to flower and been mowed aggressively in late summer- this is county policy. In the west, leafy spurge is a hugely troublesome invader, covering thousands of acres of rangeland. Tordon, a highly toxic herbicide, is the recommended agent. Ranchers have been working to use sheep (who find the spurge palatable) to manage the undesirable plants and prevent them from spreading. It is the hope of the editors that over the coming decades such nuanced, thoughtful and holistic management might be possible over greater acreages in the United States, as the numbers of land managers, farmers, growers and observers grows in number. - SvTF

Know Your Poisonous Plants

Death Camas and False As-phodel—Poisonous plant to live stock in northern Rocky Mountains and the Pacific Coast States and Canada.

False Hellebore—Sheep and other young animals eat the leaves and chickens may eat the green tops or seeds with fatal results.

Hemp—Produces a powerful narcotic drug and may cause death. Grown as a crop in Central and Midwestern States. Grows as weed in waste places.

Woolly Locoweed—Flowers violet purple. Very poisonous to live stock. From South Dakota to Mexico, west to Wyoming, Colorado and Arizona.

Crazy Weed—Flowers purple, violet or white. Another form of locoweed with same effect on live stock. Found in western Minnesota to Montana, Arizona and Mexico.

Geiger's and Menzies Larkspur—All larkspur are more or less poisonous. The first—Montana, Wyoming and Colorado. The second—west Canada to New Mexico and California.

Corn Cockle — Common weed in wheat fields. It is poisonous to farm animals and in poor wheat flour to humans.

Wild Black Cherry—The leaf especially when half-dry is a serious poison. It is found in a variety of forms all over the United States.

Sneezeweed—Poisonous to sheep, cattle and horses, in some cases it causes death.

SOME POISONOUS PLANTS OF NORTH AMERICA

Some plants are poisonous to some persons, while not harmful to others, whether they are touched or eaten. So are plants eaten by farm animals which cause various symptoms of sickness or death. Other plants will effect the condition of the milk if eaten by cows, therefore dangerous to the person drinking it.

Poison Ivy—Causes skin irritation and blistering by touching. Grows as bush or climbing vine. Eastern and central states.

Oak leaf Poison Ivy—Is found from New Jersey, Delaware, Virginia south and southwest.

Poison Oak—Same effect as poison ivy. It is found in the Pacific coast region and occurs as a bush 4 to 8 ft. high.

Poison Sumach—Same effect as poison ivy, but more severe. It grows only on wet ground, but sometimes along the road and borders of swamps.

Poison Hemlock—A deadly poison when eaten by men or animals. It grows in waste places from 2 to 7 ft. high. Often around farm buildings.

Water Hemlock—Is the most violent poisonous species in the United States for men and animals.

Poke Weed—The roots are poisonous also the fruit. Although the young leaves are used for food when cooked. It is common in Eastern United States.

White Snake Root—Is poisonous to cattle and sheep. The principle cause of "trembles" and "milk sickness" and milk sickness in human beings.

Jimson Weed—Is poisonous to humans and animals. Children sometimes eat the berries. Contact with the plant may cause skin eruption to some persons.

and suddenly an exhalation
corneas whet, slept roots illuminate
light, the young lung of
anticipation, convection. winter's
weep renews

–Leonora Zoninsein

APRIL

———

DRIZZLE

APRIL 2013

1 1841–Brook Farm, a utopian farming community is founded in Massachusetts. It's primary appeal was to young Bostonians who shrink from the materialism of American life, & the community was a refuge for dozens of transcendentalists, including Ralph Waldo Emerson & Nathanial Hawthorne.
M ♏ ♄ ☉ 6:39 19:21

2 TU ♐ ♄ ☉ 6:38 19:22

3 W ♐ ♄ ☉ 6:36 19:23

4 TH ♐ ♑ ♄ ☉ 6:34 19:24

5 Today in Tibet Lamas celebrate the SUNNING OF THE BUDDHA. by bringing Buddha statues out of temples of abstract tranquility to enjoy the sun.
F ♑ ♄ ☉ 6:33 19:25

6 SA ♑ ♒ ♄ ☉ 6:31 19:26

7 SU ♒ ♄ ☉ 6:30 19:27

8 M ♒ ♄ ☉ 6:28 19:28

9 TU ♓ ♄ ☉ 6:26 19:29

10 NEW W ♓ ♄ ☉ 6:25 19:30

11 TH ♈ ♄ ☉ 6:23 19:31

12 F ♈ ♉ ♄ ☉ 6:22 19:33

13 SA ♉ ♄ ☉ 6:20 19:34

14 In 1930, Police arrest over 100 Chicano farm workers for their union activities in Imperial Valley, California. Eight will be convicted of so-called "criminal syndicalism."
SU ♉ ♄ ☉ 6:19 19:35

15 1452: Leonardo da Vinci is born
AG M ♉ ♊ ♄ ☉ 6:17 19:36

16 TU ♊ ♄ ☉ 6:15 19:38

17 W ♊ ♄ ☉ 6:14 19:38

18 TH ♊ ♋ ♄ ☉ 6:12 19:39

19 F ♋ ♄ ☉ 6:11 19:40

20 Rice Planting Day in Thailand
SA ♌ ♈ ☉ 6:10 19:41

21 SU ♌ ♈ ☉ 6:08 19:42

22 M ♌ ♍ ♈ ☉ 6:07 19:43

23 Something stupendous takes place on the earth as a result of the full moon's forces. These forces shoot into all the vegetative growth; but only if the full moon was preceded by some rainy days.
TU ♍ ♈ ☉ 6:05 19:44

24 W ♍ ♈ ☉ 6:04 19:45

25 The time period 12 hours before and after a solar or lunar eclipse are unfavorable for sowing seeds. A partial lunar eclipse occurs tonight, though it will not be visible in North America.
FULL TH ♍ ♎ ♈ ☉ 6:02 19:46

26 F ♎ ♈ ☉ 6:01 19:47

27 PG SA ♎ ♏ ♈ ☉ 6:00 19:48

28 SU ♏ ♈ ☉ 5:58 19:49

29 M ♏ ♐ ♈ ☉ 5:57 19:50

30 in 1895 French-Canadian doctor-novelist, Philippe Panneton, who portrays people caught in the transition from primitive rural to modern urban life, is born in Trois-Rivières, Quebec. He wrote Trente Arpents (Thirty Acres, 1938) about the plight of a small French-Canadian farmer forced by economic & social upheavals of the late 19th / early 20th centuries to migrate to the city.
TU ♐ ♈ ☉ 5:56 19:51

zodiac signs

Cancer	Scorpio	Pisces
♋ Good for planting, irrigating, grafting · transplanting	♏ Exceptional for vine-type growth	♓ Good for root crops but a very poor time to make manifest
Most Fruitful	Fruitful	Fruitful

Taurus	Capricorn	Libra
♉ A time to plant lettuce, cabbage, sturdy stalks · root crops	♑ Good time to plant tubers · root crops	♎ Time for seeding of hay, grain, flowers and crops where beauty is concerned
Semi-fruitful	Semi-fruitful	Semi-fruitful

Aries	Sagittarius	Aquarius
♈ Time for cultivating · planting where weed-destruction is sought after. Chores should now will be done	♐ Best for destroying unwanted plant life	♒ Suitable for weeding · destroying of unwanted growth
Semi-barren	Semi-barren	Barren

Virgo	Gemini	Leo
♍ Good for bringing order to the garden · general cleanup	♊ Great for cultivation · destroying unwanted plant life	♌ A time for killing weeds · unwanted growth. Celer should be drawn off the orange—& good time to cut hay—if only hay is desired on hair in the increase of the moon.

ADVICE
FOR BUSINESS MEN

Don't wait for things to happen. Make em' happen.
Don't think you know it all. You don't.
Don't wait for a thing to be successful. Pitch in and help.
Don't expect too much. Then you won't be disappointed.
Don't wait for tomorrow. Tomorrow never comes.

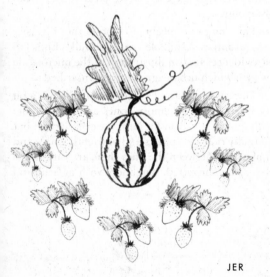

JER

Organic Power

by PETER MAURIN
SUMMARIZING
GEORGE BOYLE, 1941

Do we stop to think that organic power was socialized by the creator? Organic Power is an external force, is everywhere and yields free for the tending. By no edict of the state does the rain ripen, or the grass send up its shoots. Organic power is freer than that, yet designed, parceled out, apportioned down, in forms beyond the counting and is useable by the commonest man. The organic power is free, and distributed power.

Sidewalk Ends
A map of a city farm, Providence, RI

by TESS BROWN LAVOIE

City agriculture at its most vital can be modeled after city geography; in fact, one way to understand urban farming is as its own city, with a set of interconnected resources and populations. The perimeter of an urban farm should be articulated broadly as an expanding net. The boundaries can overlay the whole city.

Below is a map of Sidewalk Ends Farm in Providence, RI. You may notice that the area of Sidewalk Ends is a small dot on this map, but it operates within an expanded perimeter. Recorded here is the network of resources that serve and are served by Sidewalk Ends and the Little City Growers Co-op. Urban farmers depend on the multi-functionality of space and material that is made possible by the abundance and proximity of resources in urban areas. A coffee shop is not just a coffee shop for example; it is a source of coffee grounds for the compost, burlap sacks used as potato planters and to line raised beds, and a venue for agricultural produce. Urban farmers work to develop new economic and social channels. When physically represented on a map, these channels illustrate the shape of urban agricultural interactions, and the routes (which we travel on our bikes, with our laden bike trailers) that we travel to connect them. This map is a map of a small city as it is articulated by its food economy, and its agriculture. The urban farms in Providence, and the many sites that serve them, are features of their own urban system.

Urban agriculture generates experimental economies. These economies are built on innovative entrepreneurship and neighborly values. Urban farmers invest themselves in their neighborhoods' food access and beautifulness (which is only sometimes a factor in city planning). We divert city waste and compost it into energy. We rely on our community's appetite to survive. Our neighborhood has told us it is hungry not only for food grown within the city limits, but for green open spaces instead of vacant lots, for eating to be a local (i.e. socially and politically located) project, and for creative and social economies.

This map is meant to show the diversity of resources available to Sidewalk Ends in Providence. It is an illustration of the micro-scale within which urban agriculture is inscribed.

There is blank space on this map, which is a bit of a fallacy; these empty zones are always quite full, just as a vacant lot farm is never vacant, but actually rich with plants and nutrients, native and imported. Between the lines there are vacant lots yet to be converted, neighbors with agricultural

JS

backgrounds or farmerly aspirations, and many beautiful and well-cared for backyards. There are all of our CSA members. Between the lines, there are thousands of people who eat food. When urban agriculture feels too micro, it takes only a little perspective to zoom out and see how much work there is to be done. Though Providence is diverse, the farmers markets are less so. Part of the work of urban farming is to make our food culture diverse and inclusive, a holistic map.

This map represents Providence's interconnected geography, and the way food growers and sellers interact with the city and its consumers. With the support of cit. governments, and the development of good municipal and federal policy, urban agriculture will expand this network. It is work to be approached slowly: lot by lot, neighbor by neighbor. When this map is a tangled web, when it is a dense and populous network, then city agriculture will be fully urban.

OUR FARM IS SIDEWALK ENDS FARM, AND WE CAN BE FOUND ON FACEBOOK, OR VIA EMAIL AT SIDEWALKENDSFARM@GMAIL.COM

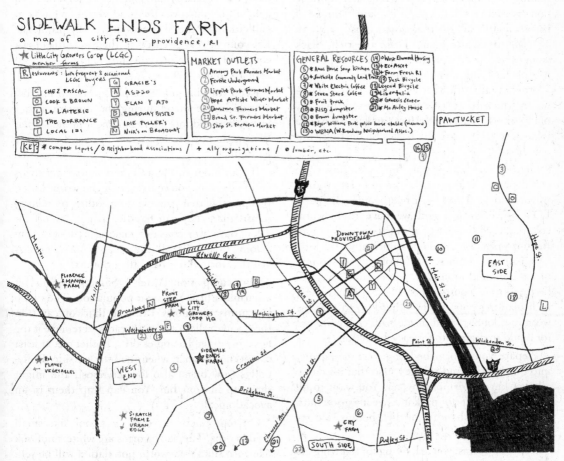

SIDEWALK ENDS FARM
a map of a city farm · providence, RI

★ Little City Growers Co-op (LCGC) member farms

R restaurants : both frequent & occasional LCGC buyers

C CHEZ PASCAL
O COOK & BROWN
L LA LAITERIE
D THE DORRANCE
I LOCAL 121

G GRACIE'S
A AS220
Y FLAN Y AJO
B BROADWAY BISTRO
F LOIE FULLER'S
N NICK'S on BROADWAY

MARKET OUTLETS
1 Armory Park Farmers Market
2 Fertile Underground
3 Lippitt Park Farmers Market
4 Hope Artiste Winter Market
21 Downtown Farmers Market
22 Broad St. Farmers Market
23 Ship St. Farmers Market

GENERAL RESOURCES
14 o West Elmwood Housing
5 ★ Amos House Soup Kitchen
6 + Southside Community Land Trust
7 ★ White Electric Coffee
8 ★ Seven Stars Coffee
9 ★ Fruit truck
10 ø RISD dumpster
11 ø Brown dumpster
12 ★ Roger Williams Park police horse stable (manure)
13 o WBNA (W. Broadway Neighborhood Assoc.)
14 o West Elmwood Housing
15 + Eco Asset
16 + Farm Fresh RI
17 ★ Dash Bicycle
18 Legend Bicycle
19 ★ Libertalia
20 ★ Genesis Center
21 ★ McAuley House

KEY: ★ compost inputs / o neighborhood associations / + ally organizations / ø lumber, etc.

PAWTUCKET

I-95

DOWNTOWN PROVIDENCE

Atwells Ave.

Manton

Valley

FLORENCE & MANTON FARM

Broadway N
FRONT STEP FARM
LITTLE CITY GROWERS COOP HQ

Knight St.

Washington St.

Westminster St.

Red PLANET VEGETABLES

WEST END

SIDEWALK ENDS FARM

Cranston St.

Bridgham St.

SCRATCH FARM & URBAN EDGE

Elmwood Ave.

CITY FARM

SOUTH SIDE

Dudley St.

EAST SIDE

Hope St.

N. Main St.

Point St.

Wickenden St.

Broad St.

Dean St.

I-195

LBL

Home Vermicomposting

by LAURA BROWN LAVOIE

Worm composting is a major component of my urban homestead. Worms consume almost all the vegetative waste that is produced in my household, and they live with us, so there's no long commute with stinky scraps. We do maintain several traditional compost bins at our farm, but I like to teach my neighbors and friends about worm bins, because I think they are great for home composters, especially in a city, where most people don't have a lot of outdoor space for maintaining a regular pile. Worm bins are odorless, don't attract rats, don't require a large volume of brown matter (as do traditional bins), and can be kept inside the home. I use the rich, black compost my worms produce to make our potting mix in the spring and add it to my beds all season. Worm bins also drain a dark brown, nitrogen-rich water, which I harvest and dilute to fertilize my plants (both watering with it, and using it as a foliar feed). Here are the basics of starting your own bin:

Get two identical opaque bins, with lids. (The size depends on how much compost you produce in your house. I live with five people who eat a LOT of vegetables, and we have three worm colonies, each the size of about three milk crates. But one worm bin should be sufficient for the food scraps of most homes.) Drill holes in the bottom of one bin, and around its upper rim. That will be the bin for your worms. Place risers (I use 4 inch planting pots) in the other bin, so that the worm bin, once placed inside, will create a space for liquid to drain into for easy collection.

Prepare your worm bedding: tear newspaper into 1-inch strips and moisten it so that it's damp all over but not dripping water. You want to fill the whole bin but not pack it. Give it a good fluff. Also throw in a handful of dirt, because worms need grit for their digestion.

Get your worms: you're looking for red wigglers (Eisenia fetida). If you have a friend with a worm bin, she probably has some worms to spare. Otherwise, you can order them online. I got mine from the Worm Ladies of Charlestown, Rhode Island. They sell them by the pound at a farmer's market here, and also ship locally, but there's probably a worm purveyor near-ish to wherever you are!

Spread your new worms across the bedding, and leave the lid off for ten minutes, so they burrow away from the light. (Worms hate the light.) In terms of feeding, start your worms off slow-- maybe just a little rotten lettuce or some squashed tomatoes in with their bedding. After you dump the worms in, cover them with a whole dampened piece of newspaper, like a blanket. In general, if you cover the food scraps this way, it will eliminate the likeliness of fruit fly problems. Let your worms get used to their new digs for a couple of days before you start loading them up with all your kitchen scraps.

Keep your bin in a cool place. My worms stay outside in the shade during the summer, and I bring them indoors to my basement in the winter. Worms do not like temperatures over 84 degrees Fahrenheit.

If your bin is stinky, add more newspaper strips and make sure you've harvested the worm water. It is possible that you might be adding more food scraps than they can handle, so just lay off for a few days and consider making another worm colony to handle all the scraps you've got. A healthy worm bin is basically odorless.

On "companion critters": Sometimes you get a harmless little red mite population. They look line miniscule rubies. I've read that you can trap them with a slice of white bread and remove it the next morning. The mites don't bother me, at least not in the summer, when the bin's outside. Rolly-pollys (sow bugs) also love compost, and might show up in your bin. You can trap them in an avocado peel.

On reproduction: worm eggs look like small golden orbs. The baby worms are white, and look like corn silk. Your worm population will be self-regulating (they will reproduce only up to the point where their population can be sustained by the space.) But, be generous with your worms! When you notice you have a lot, find a friend or neighbor who doesn't have them yet.

Harvesting the worm soil is not easy, but it can be fun (especially if you have a bunch of kids around to pick out the worms!) If you don't have happy worm-pickers, you have a couple of options. If your bin is full, you can take out half its contents, worms and all, and spread it on your garden, while leaving the remaining half to maintain your population. After all, there's no harm in adding worms to the garden. If you're inclined to save all your worms, take advantage of their hatred of light. Dump your whole bin on a tarp, and make little cone-shaped mounds. Go from mound to mound harvesting off the top parts where the worms have burrowed away.

Continue this process until you have mostly piles of worms, which you can throw back in the bin to start anew.

Oh! I almost forgot: naming your worms. There are around 1200 red wigglers in a pound of worms. I've found it difficult to tell them apart, so I usually give them one name. My bins have worms named Boris, Laura, and Mary Woolstonecraft.

I am indebted in my worm composting practice to the excellent compost handbook The Composting Cookbook by Karen Overgaard and Tony Novembre. They also recommend Worms Eat My Garbage by Mary Appelhof. The Worm Ladies of Charlestown have a website, and I've found that they will answer my questions via email, with great detail and generous expertise. Happy Worming!

OUR FARM IS SIDEWALK ENDS FARM, AND WE CAN BE FOUND ON FACEBOOK, OR VIA EMAIL AT SIDEWALKENDSFARM@GMAIL.COM

Pomona

ROMAN GODDESS OF THE ORCHARD

GM

Manifesto

FRIENDS OF THE LAND

A NON-PROFIT, NON-PARTISAN ASSOCIATION FOR THE CONSERVATION
OF SOIL, RAIN AND MAN

I. *Evidence of the Need*

. . . A good land; a land of brooks of water, of fountains and depths that spring out of the valleys and hills; a land of wheat and barley, of vines and fig trees and pomegranates; a land of olive oil and honey . . . Here thou shalt eat bread without scarceness; thou shalt not lack anything. . . .

Deuteronomy, 8, 7–9.

The waters wear the stones; thou washest away the things that grow out of the dust of the earth; and thou destroyest the hope of man. . . . If my land cry against me, or that the furrows thereof likewise complain . . . let thistles grow instead of wheat, and cockle instead of barley. The words of Job are ended.

The Book of Job.

IT IS AN OLD STORY, often repeated in the time of Man. We have talked a lot about it in this country lately. We have barely begun to do something about it in a large, sensible and connected way.

The need to do more is urgent. The record is plain. Over vast areas we stand confronted with defaced landscapes, depleted water supplies, grave dislocations in the hydrologic cycle, and an all but catastrophic degradation of soil and Man.

We have hurt our land. We have made much of it ugly in the plain implication that land laid to waste will not support that measure of individual freedom and those constantly higher standards of living which we as Americans have been led to expect.

Down our streams every year go enormous quantities of plant food elements—nitrogen, phosphorus, and potash that might have produced bread, meat, milk, and garments. This huge loss represents only part of the annual erosion bill. Erosion not only removes plant nutrients; it carries away at one disastrous stroke the available plant food, the material from which plant food is made, the micro-organisms that aid in the manufacture of available plant nutrients, the mineral matter that holds these organic and inorganic materials—the whole body of the soil.

Soil misuse makes people poor. Soil displacement is followed by human displacement. The first shock of displacement is felt in the open country. But soon, as yields and trade fall off, it is also felt in the towns.

Any land is all of one body. If one part is skinned, bared to the beat of the weather, wounded, not only the winds spread the trouble, dramatically, but the surface veins and arteries of the nation, its streams and rivers, bear ill. Soiled water depletes soil, exhausts underground and surface water supplies, raises flood levels, dispossesses shore and upland birds and animals from their accustomed haunts, chokes game-fish, diminishes shoreline seafood, clogs harbors, and stops with grit and boulders the purr of dynamos.

Eroded soil is soil in some part dead, devitalized. Soil debility, soon repeated in nutritive deficiencies, spreads undernourishment. Evidence on this point is far from complete; but the trend of accumulating findings is unmistakable. If the soil does not have it in it, plants that grow there do not; nor do the animals that eat those plants;

THE LAND

nor the people throughout a country who eat those plants and animals. Soil debility soon removes stiffening lime from the national backbone, lowers the beat and vigor of the national bloodstream, and leads to a devitalized society.

We, too, are all of one body. We all live on, or from, the soil.

No matter which political party gains ascendancy as the years go by; whether the swing be from middle Left to far Right, or to the farther Left; whether we remain at peace or go to war again, this fact will remain: so long as we keep on scrubbing off, blowing off, killing off our topsoil, business and social conditions in this country will remain fundamentally unsound.

II. A Statement of Purpose

WE THEREFORE NOW INTEND to organize and to bring quickly into action a non-profit association or society to support, increase and, to a greater degree, unify, all efforts for the conservation of soil, rain and all the living products, especially Man.

III. What We Can Do

WE INTEND first to work with friends of conservation, both lay and professional, here in this country, and later with like-minded men and women in other lands.

With the conservation idea advancing to a wider outlook and more practical techniques of research and husbandry; with conservation becoming, in effect, a working philosophy to reconcile the ways of Man and Nature—the time is right for such a society to form and act.

The need is imminent. Much of the civilized world is at war again, sick at heart and weary. Even this far removed from the main centers of pressure on soil and humans, we feel, and shall continue to feel, the strain and tension. A wartime psychology fixes attention on devices of slaughter and destruction. It diverts human effort and ingenuity from studies and devices to perpetuate the source values of humankind.

All friends of conservation need now to move and speak out together as never before.

Whether your principal personal interest be in soil, grass, trees, songbirds, game, flowers, livestock, landscape or outdoor recreation; and whatever your occupation—farmer, banker, forester, agrostologist, journalist, anthropologist, ecologist, teacher, student, or what not—we can all work together for the good of the land.

We may promote the conservation of land and water resources in the United States of America by:

1. Assembling information regarding the economic, industrial and social need for conserving our land and waters; placing before the people of the country various issues and problems in land and water conservation; and forwarding in the interests of the public, specific policies of conservation.

2. Encouraging the organization of affiliated regional and local groups.

3. Preparing and making available to our membership and to the general public a magazine, THE LAND, and other literature on the technique, importance, and significance of land and water conservation, and recommending to our members suitable literature prepared by other organizations, operating in special fields.

4. Fostering investigation, exploration, research and experimentation into the science of soil and water

MANIFESTO

conservation, and recognizing achievements in this field by electing outstanding scientists as honorary members of the society, or by special awards.

5. Encouraging and furthering the practice of land and water conservation by individuals, cooperative groups, States and subdivisions thereof, and the Federal Government, and promoting legislative measures and the efficient and economical use of public funds in furtherance thereof.

6. Recognizing outstanding accomplishments in land and water conservation by farmers, soil conservation districts and other local groups, by suitable citations and awards.

7. Promoting inclusion in the curricula of our educational systems of courses on the significance and technique of land and water conservation.

8. Fostering the participation of the youth and youth organizations and especially unemployed youth, in a moral equivalent of war against wastage of soil and water.

9. Cooperating with other organizations interested in the conservation of trees, grass, wildlife and people in promoting common objectives.

10. Convening periodical conferences in various parts of the country to obtain wider recognition that soil wastage threatens our institutions.

We could promote the conservation of land and water resources in this and foreign lands by:

1. Appointing in each foreign country a "correspondent" (without pay) for the exchange of information.

2. Maintaining a clearing house of information on conservation in foreign countries, and from time to time publishing a survey of foreign activities in this field.

3. Encouraging the establishment in this and in foreign countries of private organizations for the furtherance of conservation.

4. Furthering the adoption of courses on conservation by educational institutions throughout the world, and the granting of traveling fellowships to foreign officials and students for the study of conservation.

5. Advising foreign governments on methods of establishing programs for land and water conservation.

6. Assisting foreign mission and educational organizations in incorporating programs for land and water conservation as a major objective of their work with foreign peoples.

7. At an appropriate time calling a World Conservation Congress.

> MORRIS LLEWELLYN COOKE,
> CHARLES W. COLLIER,
> BRYCE C. BROWNING,
> CHARLES E. HOLZER,
> RUSSELL LORD.

THE LAND

A New Anthology About Old Ideas

by HENRY TARMY

The dawn of World War II would seem an inauspicious time to launch an organization dedicated to agricultural sustainability. Soil degradation is, by nature, a quiet and insidious menace. Erosion doesn't lend itself to sound bites or excite the passions of the populace quite as much as, say, Germany annexing Poland. But in March of 1940, with war on the horizon, a diverse group of interested citizens founded a society they called Friends of the Land, 'a non-profit, non-partisan association for the conservation of soil, rain and man'. Their mission was to combat the societal inertia surrounding matters of conservation and farming, with a campaign designed to educate the public and to catalyze a movement.

Friends of the Land recognized that while 'direct human killing is bound to deflect attention from the slower tragedy of soil killing... the need of doing the sort of work and teaching that we set out to do is even greater in times of world strain and confusion than in times of peace and ease...' The group wanted to illustrate for city and country dwellers alike certain inescapable connections: the fundamental importance of the soil AS A resource for the health of the nation and of the individual. In their founding manifesto, the friends group wrote that,

"The need to do more is urgent. The record is plain. Over vast areas we stand confronted with defaced landscapes, depleted water supplies, grave dislocations in the hydrologic cycle, and an all but catastrophic degradation of soil and Man."

To promote their message of ecological literacy they used an innovative and diverse set of communication tools: radio shows, movies, conferences, conservation camps, farm 'make overs' and other media.

The Land, their quarterly magazine, was at the heart of this effort from ITS FOUNDING IN 1941 until its final issue in 1954.

In a 1941 essay for The Land, Aldo Leopold wrote,

"Conservation, I fear, is still in large degree a parlor game. We tilt windmills in its behalf in convention halls and editorial offices, but we seldom discuss the concrete realistic problems of actual land-use except for purposes of shocking an apathetic city audience.... A [difficult] task has been laid upon the coming generation... How to join the life of a local community without 'going native' intellectually; how to muster courage to unravel land-use problems which are, at best, only partly soluble; ...these are tasks indeed! It becomes increasingly clear that a new breed of cats must own the land before much headway is possible."

In 2013 the continued relevance of their commentary is at once wonderful and tragic. It is both an inspiration and a testament to our failure to address the fundamental problems they articulated over seventy years ago.

For us as inheritors of the Friends of the Land's vision for an agricultural system that builds soils and communities rather than destroying them, this current surge of popular interest in our farms and food systems is a hopeful sign: a new breed of cats has arrived. A rapidly expanding group of mostly young new farmers is out on the land taking practical steps to grow a better future. It falls on us to ensure that these new cats gain traction and prominence-- that the importance of considering the ecological whole is brought from the margins to the mainstream.

FOR MORE INFORMATION ON THIS REVIVAL AND THE WORK OF FRIENDS OF THE LAND PLEASE READ THE ANTHOLOGY OF THE LAND, SUPPLEMENTED WITH ESSAYS FROM CONTEMPORARY THINKERS, FORTHCOMING FROM THE UNIVERSITY PRESS OF KENTUCKY.

Rerooting
The Motor City

by AMANDA MATLES

A conversation between urban agriculture leaders in Detroit, MI and the Paper Tiger Video Collective, New York, NY. This is an excerpt of a PTTV documentary filmed in Detroit from 2010-2012. Images here are from the documentary.

PTTV: When people see stories about Detroit and its regeneration in the news, what are the mainstream stories being told?

Lottie Spady: So basically the mainstream narrative around Detroit is that it was a "barren wasteland" that was "empty" and "destitute" and going "down the tubes", now there are people moving in, usually shown as young white people, at an alarming rate to homestead and hook up shop, or set up a farm, or whatever, and largely, it's not being done in an effort to retain or maintain the communities that were already there. The whole urban agriculture movement is being glamorized and popularized in the media as some class based, you know, hippie, yuppie thing, that is new. It is just not true.

PTTV: So if the new urban "frontierists" didn't start the urban farming movement in Detroit, who are the real originators of Detroit's urban agriculture networks?

Shea Howell: The initiation of this urban garden movement started more than thirty years ago, primarily by older African American women, who had roots in the American south, who called themselves the Gardening Angels. Their work was about community memory, and about beauty and about how to reclaim a place through meaning. So many people and businesses were fleeing the city. They would collect the peach tree you planted for your daughter's birth, or the roses you had growing around the porch and take care of them in the empty lot gardens of burned out blocks. But their care was also about food. Because Detroit has officially been a food desert and people needed to eat.

PTTV: What are some of the groups spearheading urban food system work in Detroit, what do they do, and what remains to be done?

LS: Grown in Detroit Cooperative has been developing a network of small growers. We are over 80 community gardens and urban farms throughout Detroit. We've sold more than 23,000 pounds of local produce in 2009 and also managed to donate 1,100 pounds of food to soup kitchens and churches. The Detroit Food Justice Task Force will be working to establish a set of Food Justice Principles that work for us here in Detroit. We need a framework of food security, food sovereignty, and food justice specifically for Detroit.

CH: D Town Farm is the agricultural project of The Detroit Black Community Food Security Network. Charity Hicks: We believe that nothing is more intimate than food. Food is everyday. So how do we scale it up in such a way that it's sustainable, and honors the earth, that it promotes economic and social justice, and it doesn't exploit anyone? Detroit has to have some self-help, some advocacy and some self-determination. And we believe and we know that once we enhance the black community here, everyone will benefit. We work on food policy, food systems, food education, and food literacy. Our model is collaborative, niche market, and small holder, we are in line with La Via Campesina, and all the small holders world wide, which are operating as a carbon seat. Low diesel, low carbon, a lot of small mechanization, and a profound sense of stewardship on land and the ecology of a place. It's not just a monoculture. And you'll see small holders with variety, biodiversity, and how they hold their land is much more natural and resilient than monocultures. We sell a lot of our produce through the Grown in Detroit Cooperative.

Patrick Crouch: I am the produce manager at Earthworks Farm. We have a Youth Farm Stand Program where young people learn urban farming and market selling. Our motto is "Making a Just Beautiful Food System for All"
We also do a big market at the soup kitchen, actually, so anybody can come, and we're able to take EBT cards and WIC vouchers, so folks can come by and get stuff that they want to take home. A lot of the food we grow goes to volunteers. A lot of it goes to the soup kitchen. We're interested in getting fresh quality food in the neighborhood, but we're also interested in using it as a community organizing tool, that it's bringing folks together, creating a common space so that people can start talking about other social justice efforts going on in Detroit.

SH: John Hantz is a guy who's made his money mostly through the financial markets. The Hantz Farms people want to do not urban gardening, but industrial farming inside the city. We are very concerned with the city's push for corporate farming. He has proposed taking several thousand acres and creating industrial farming in an enclosed method of some sort. The issues that it raises are that high tech industrial farming does not employ a lot of people. Secondly, nobody has had a high level industrial farm this close to a major water source. It's been very controversial.

LS: The other thing with Hantz farms is that we don't know if this is proposed to be an organic farm or at least pesticide free. How are they intending to manage that piece of it, and no one was talking about doing it in such a way that it does not further contribute to environmental injustices as it is. They seem to be accountable to no one.

PTTV: What are some of the key issues that drive Detroit's agricultural movement forward?

Kwamena Mensah: In a city that's 90% black that has very little representation in the food system other than like, a store clerk... The Detroit Black Community Food Security Network is a statement of resisting. We can do something better than this. This is a system of self-determination, an example of resistance. Resistance to the status quo, resistance to being disenfranchised; this is a key issue.

CH: Regular grocery stores have left! The last one left about three years ago. We are officially a food desert. We have very little retail options here in Detroit... we have a lot of markets, like Eastern Market, but grocery stores? It's a real problem here. So what do we do? We start canning classes, and cooking classes, and recipe sharing. Spice industries, local bakeries and things like that because it's essential. We've got a generation or two of people who have forgotten how to cook because they were so busy working in the factories. So this is an opportunity, all of these community classes and economic initiatives, all of this can be done because of this urban ag movement.

Nefer Ra Barber: We're trying to get the word out to everybody about the Urban Agriculture Work Group. We're working on changing government policy and codes. We are exploring chickens and hoop houses and goats and beekeeping and rooftop gardening and all the aspects of urban agriculture. We have to allow for that within our zoning and within the city's local ordinances. What's the impact on the adjacent land uses, and the whole political ramifications of it all? We're so far successful on some fronts, but it's a struggle to convince people that this will be beneficial. Also, not being zones doesn't stop anyone from farming illegally. But it would accelerate the small farm movement if we were zoned.

LS: If we had a number of small scale urban growers increase a little, as opposed to one large scale grower, we would be able to put that money into those individual household economies as opposed to one corporate pocket.

PTTV: How do you see these Detroit struggles for environmental and economic justice related to worldwide struggles?

Kadiri Dobey: Cut dry, the world food system is racist. Pretty much everything is. There is a racial wage gap in all four food sectors. Black people have very little representation in the food system. A lot of the groups here are part of the Undoing Racism workshops to try to address racism in our communities and way of organizing our economies. D town farms a lot of the members there are responsible for spearheading the food policy council issues to not just reform but to actually change this food system. Its not enough to just localize it. For us to have a just food system, we have to address the racial problems and issues that are a systemic part of that food system.

KM: There's a land grab going on in Detroit right now. There's land grabs all over the planet, particularly in Africa. Sates are rushing to grab thousands and ten thousands of hectares for mineral rights, for mining, just for everything - and particularly for agriculture. There's profound speculation on agricultural land world wide. It's a food crisis. So what you are seeing over in Africa, you are seeing here in some of these inner cities like Detroit. People are coming in and trying to purchase - or the city will just give them - 300 or more acres of land to start large scale

commercial farming in the city. As if that's going to bring a bunch of jobs back. Those operations are only promising 20-30 jobs.

LS: If one person or entity owns a bunch of land and the folks that work on it are just employees, if they are not going to be a workers' cooperative, or something like that, then how is that anything other than creating a modern day plantation? So lets just call it what it is!

PC: Detroit's pretty much the biggest boom town there ever was. Giant business got us into this mess in the first place and I don't see how its going to help us to have an economy based on a single industrial exports, whether that be agriculture or the auto industry. Ultimately I want to see a Detroit that is a self-sufficient economy, first and foremost an economy based upon our needs here. So we need to think about building all our own tools and our power and all the things that we need to thrive. And the access to land by those who actually live in the city is still grossly disproportionate. A lot of the land is still owned by those who live outside the city with other interests, and so I think that's part of the reason why we have the situation we're in now. It's because folks haven't had control over the land they live on. So what is more equitable, one person owning 10,000 acres, or 10,000 people collectively owning 10,000 acres?

LS: In Detroit there has been an intentional divestment of all resources until everything dropped to such a low, then the desired populations can move in and scoop everything up cheaply and effectively force everyone out who can't afford to buy back land. Now's the time to root ourselves in a way that we can't be so easily displaced again - and recognizing our value! Before it's told to us through the media, urban removal, or displacement.

CH: People world wide are waking up to the local food system. People are creating local vibrant economies. It's a world wide movement. You have to think global but your real actions have to be local. So even though we have this D Town Farm here in Detroit, we're connected with farmers in Africa, and Europe, and South America, in Haiti and Havana, and direct reaction to the greedy bastard syndrome is the resiliency of hu-

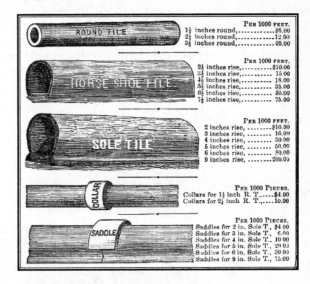

EDITORIAL NOTE: We talk about "field tilings." This is a practice of land improvement for modern farming. Tiles move water swiftly to bring logged soils into cultivation, to prevent their becoming "water logged" and soggy, which can create poor growing conditions as well as compacted soil and damage to the agricultural potential of the land over time. The practice of soil drainage and subsoil tiles was practiced early in our history by progressive farmers, but found widespread application in the soil conservation bonanza of the late 1930's, when national response to the dust bowl was strongest. Early tiles were made from ceramic, but today's tiles are plastic, sited with lasers, and installed using ditch-diggers and backhoes. The millions of miles of tile under the soils of our productive agricultural regions also shunt agricultural chemicals and soil runoff directly into streams and ditches, instead of seeping slowly into the ground table, or via the microbe-heavy filter of a swamp or marshland. Intensive management of our agro-ecosystems in a way that ignores 'other users' such as wildlife is typical of this american empire. It is thanks to tiles that we have such a uniform agricultural landscape, allowing monocultures on such a massive scale.
- SvTF

'Quit thinking about decent land use as solely an economic problem, but examine each question in terms of what is ethically and aesthetically right, as well as what is economically expedient. A thing is right when it tends to preserve the integrity, stability and beauty of the biotic community. It is wrong when it tends otherwise'.

– ALDO LEOPOLD, ON LAND ETHIC

· ·

"Until one is committed, there is hesitancy, the chance to draw back-- Concerning all acts of initiative (and creation), there is one elementary truth that ignorance of which kills countless ideas and splendid plans: that the moment one definitely commits oneself, then Providence moves too. All sorts of things occur to help one that would never otherwise have occurred. A whole stream of events issues from the decision, raising in one's favor all manner of unforeseen incidents and meetings and material assistance, which no man could have dreamed would have come his way. Whatever you can do, or dream you can do, begin it. Boldness has genius, power, and magic in it. Begin it now."

– JOHANN WOLFGANG VON GOETHE

BEFORE

On Courage And Resistance

The timeless Oscar Romero Award keynote address Susan Sontag delivered on March 30, 2003, originally published in the 2007 posthumous anthology At the Same Time: Essays and Speeches (public library).

Fear binds people together. And fear disperses them. Courage inspires communities: the courage of an example – for courage is as contagious as fear. But courage, certain kinds of courage, can also isolate the brave.

The perennial destiny of principles: while everyone professes to have them, they are likely to be sacrificed when they become inconveniencing. Generally a moral principle is something that puts one at variance with accepted practice. And that variance has consequences, sometimes unpleasant consequences, as the community takes its revenge on those who challenge its contradictions – who want a society actually to uphold the principles it professes to defend.

The standard that a society should actually embody its own professed principles is a utopian one, in the sense that moral principles contradict the way things really are – and always will be. How things really are – and always will be – is neither all evil nor all good but deficient, inconsistent, inferior. Principles invite us to do something about the morass of contradictions in which we function morally. Principles invite us to clean up our act, to become intolerant of moral laxity and compromise and cowardice and the turning away from what is upsetting: that secret gnawing of the heart that tells us that what we are doing is not right, and so counsels us that we'd be better off just not thinking about it.

The cry of the anti principled: 'I'm doing the best I can.' The best given the circumstances, of course.

AFTER

NC

A lot in Portland, Oregon, transformed in 7 months, by amateurs.

" A fair reward for toil,
A free and open field,
An honest share for wife and home,
of what your harvests yield."

"Stay on the farm, boys, stay on the farm,
Though profits come in rather slow.
Stay on the farm, boys, stay on the farm;
Don't be in a hurrry to go."

THESE SONGS WERE FOUND
IN A 1920 GRANGE SONGBOOK,
RE-RECOREDED BY THE CONSPIRACY OF BEARDS,
AND IS INCLUDED IN THE ACCOMPANYING
NEW FARMERS AUDIO ALMANAC
WWW.NEWFARMERSALMANAC.ORG

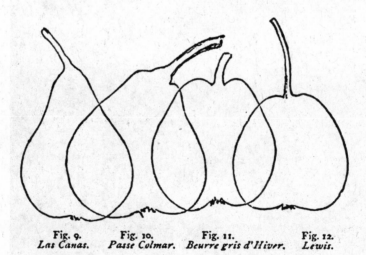

Fig. 9.   Fig. 10.   Fig. 11.   Fig. 12.
Las Canas.   *Passe Colmar.*   *Beurre gris d'Hiver.*   *Lewis.*

Community

by KRISTEN KIMBALL

One of our first seasons we lost a dozen turkey poults to a bloodthirsty raccoon. It pried the brooder door open with its nimble little hands. Raccoon kills are the messiest kind; you can read panic and depravity in the still life left behind. I remember feeling dumbfounded at the raccoon's cleverness, sad for the turkey's suffering and guilty for my failure to protect the helpless creatures. I was clearing out the bodies when one of our neighbors pulled up and took in the scene. "Well, you ain't gonna have livestock without having some dead stock," he said. Nine years in, that rings true in my bones. There is no livestock without dead stock. No harvest without blight. No joy without frustration. We try to cultivate a good balance, and when things tip the wrong way, we try to take it. There are good years and then there are the other kind. The best counterweight to trouble at those times, I think, is community. I believe that if you cultivate your community with loving care, grow deep strong roots into it, it will give back to you when you need it.

The last few seasons the weather has been strange and difficult. There was a year of constant, wearying rain, then one bookended by floods. This past year was so dry that fields all across the country cracked; the newspapers from coast to coast ran photos of farmers in their drought-stricken corn. But our farm is low and the water table is high and it was those two wet years that almost killed us. We had a new baby and a toddler, and the farm was growing too fast for its own good.

We were in a particularly bad patch two Augusts ago. The fields were swampy after a summer of rain, and the best-looking plants on the farm were the weeds. Mark was dealing with a back injury, I was dealing with two young children, and most of our employees were both green and disgruntled. One particularly bad day Mark was walking the fields by himself, making a list of what needed to be done, when a neighbor pulled

up in her car. "Get in!" she called to him. It was hot and she had the air conditioning cranked and he sank into it like a pool. She handed him a cold soda, and then got right to the point. She said she'd been watching our farm, which sits at the edge of a town of 700 people. She said that in all her years in this community she had never seen anything transform the place so much for the better. She also saw that our fields were always too wet, and that this was holding us back, and she wanted to help. She handed him a check, a big one, to drain the fields.

It was this kindness that lifted us over the rocks of a year that could have scuttled us. We started with her money and added our own and now have 50 acres drained. Fifteen of them were planted to rye last season, on ground that had never yielded anything but muddy pasterns in the past. I took the children to it at the dry height of last summer, tried to tell them the story of how it came to be; they looked at me uncomprehending, then disappeared into the rye, with only their voices to tell me where they were. I lay down and watched the golden heads, heavy with the gift of grain, wave gently between me and the bright blue sky.

Copper plate prints borrowed and printed from Rogersons Hardware Store, Hudson, NY

The Disconnect

by ANASTASIA RABIN

Too many people are not spending time engaged with the land; outside, exposed, working, digging, using their senses, and experiencing direct feedback about the way that they move, manipulate their bodies and their environment. Machines and people who are treated worse than machines, are now doing the work that feeds us.

Underpaid immigrants destroying their families to come here, and destroying their bodies when they get here. Because the cheap food, cheap lumber, cheap everything that we are demanding, taking for granted, and then oftentimes WASTING, cannot be produced at this level on a human scale OR within the natural limitations of what nature provides.

What I learned from years of the repetitive, "back-breaking" work that I did is more valuable than anything I could have ever learned in school. School teaches a lot of things that you have to un-learn in order to actually be effective in the world. Don't get me wrong, it has a lot to offer, I did learn a lot of worthwhile things and it certainly speeds the learning process when you have that structure. I am better off for it, but I will always be an advocate for the point of view of the person that spends their time working on the land. We should be listening to the people who are out there every day ON THE LAND learning experientially, and second guessing the people who get their knowledge from books and the study of theories, not the other way around.

Migrant workers, ranch hands, farmers, and the like are grossly under respected for their knowledge, their perspective, and their labor. I can't tell you how many times as a tree planter, I was told to do something that made absolutely no sense. To plant in a place that was near impossible to get to where the trees were sure to die anyways. We had to do it because someone in an office looking at a map decided, and the guy supervising us in the field couldn't or wouldn't do anything about it. When you are humping 2/3 of your own weight in tree seedlings up the side of a steep mountain at 9,000 feet elevation that shit HURTS!

Farming, ranching, and restoration work is too hard as it is. We can't afford to make it any harder with attitudes and bureaucracies like the ones we've got. A huge shift is needed if we are going to produce food sustainably from the land, and heal the damage that has been done. It needs to happen from a place of deep respect. So much we need to do while the country goes spinning off out of control in the opposite direction.

THE END OF A WAY OF LIFE. The work that did 10, 15, almost 20 years ago is not an option like it once was. I feel like I squeezed in and had those experiences when it was already obscure, and essentially coming to an end. Gas was cheap. You could live out of an old truck and get from job to job without going broke. Many companies, farms, and government agencies were happy to replace the roving bands of freaks and outlaws with machines or Mexican crews run by guys that keep them all in line. Why would you want a bunch of ruffians camped out in the woods with their 2 or more dogs for every crew member, drinking around the campfire, getting their crippled trucks stuck in the road, going into town and raising hell on payday, and making fun of the boss-man or the inspector relentlessly, when you could have a seemingly organized crew of immigrants who are all paranoid about getting deported, working twice as hard for half the money, and being shuttled in and out of the worksite from a motel?

Where can a young person go to be a part of something like that? Maybe the fishing scene in Alaska, but I have never been there to say for sure. I know that there is work on organic farms, maybe ranches too, but it's not the same. You are likely to find people there who share somewhat similar ideologies, who stay for a longer period of time and have a more intimate relationship with that place. That's all good, and learning those things is important, but the best part about what I did (for me) was that it was NOT that way. I felt very free. I loved the sort of anonymous camaraderie of sharing grueling work with a bunch of strange people for a short period of time. The reckless friendships that would form.

I never wanted to stop. I had to because it wasn't an option anymore. Not for me, not for a lot of people.

Many of the people that I worked with wanted to save $ to buy land and be self-sufficient from it, to farm essentially. That seeming to be the only other way to live without compromising their independence and individuality. I think many of them did end up with places of their own. I hear about them, but am not in touch with many people to know how it is going. We probably share many of the same heartaches.

It's a life full of irony to want to travel, wander, and be free, and simultaneously to want to settle down and farm.

OLD AND HOMELY PROVERBS FOR EVERY DAY IN THE MONTH.

1. Get thy spindle and thy distaff ready, and God will send thee flax.

2. Better ride an ass that carries us than a horse that throws us.

3. Everything comes in time to him who can wait.

4. Love rules without a sword.

5. Trust thyself only, and another shall not betray thee.

6. Nothing is lost on a journey by stopping to pray or to feed your horse.

7. Every vicious indulgence must be paid for cent per cent.

8. Better to be alone than in bad company.

9. To say little and perform much is noble.

10. Every man thinks his own geese are swans.

11. Circumstances alter cases: the straightest stick appears crooked in water.

12. An honest man is none the worse because a dog barks at him.

13. When you are an anvil, bear; when you are a hammer, strike.

14. He laughs best who laughs last.

15. He that can't paint must grind colors.

16. Wise distrust is the parent of security.

17. Idleness is the sepulchre of a living man.

18. The devil tempts all men; but the idle man tempts the devil.

19. Business is the salt of life.

20. Never measure other people's corn by your own bushel.

21. He who spares vice wrongs virtue.

22. Despatch is the soul of business, and method is the soul of despatch.

23. Like plays best with like: when the crane attempted to dance with the horse, she got broken legs.

24. A full vessel must be carried carefully.

25. That is often lost in an hour which costs a lifetime.

26. Keep yourself from opportunities, and God will keep you from sins.

27. The pitcher that goes often to the well gets broken at last.

28. Give a rogue an inch, and he will take an ell.

29. Many a cow stands often in the green meadow, and looks wistfully at the barren heath.

30. A handful of common sense is worth a bushel of learning.

31. The fire should burn brightest on one's own hearth.

Food Policy Councils:
A Look Back at 2012

prepared by MARK WINNE ASSOCIATES

This past year marked several growth spurts and critical shifts in the world of food policy councils. First and foremost was the incredible leap forward from 111 active North American food policy councils in 2010 to 193, as indicated by the May 2012 census conducted by the Community Food Security Coalition.

This jump in city, county, state, and tribal level councils sends several important signals to policy makers and food system activists. Perhaps the most obvious one is that citizens and stakeholders want a bigger role in shaping the direction of their food systems through the policy making process. Just as important, food policy is occupying ever more "real estate" on the radar screens of non-federal policymakers like city mayors, county commissioners, and state political leaders. At least at the local and state levels, there is an evident surge in food democracy.

The second major FPC-related event was the closing of program operations at the Community Food Security Coalition this past August. Unfortunately, this ended any formal organizational support for food policy council capacity building. As both CFSC's former food policy council program director and the principal with Mark Winne Associates, I have been attempting to maintain assistance for North America's food policy council movement. Being a single staff person with no funding dedicated to that task, of course, has its limitations. While I have been able to offer a platform for some of CFSC's FPC services like the FPC listserv, the North American Food Policy Council Directory, and several food policy resource documents that were in the pipeline at the time of CFSC's demise, I have not been able to respond as fully as I would like – and as I think is necessary – to the many requests for assistance. The good news, which I cannot yet say too much about, is that I'm optimistic that an organizational partner who can shore up the capacity building needs of the FPC community is on the horizon.

In spite of limited resources, however, Mark Winne Associates has provided FPC development support to a number of communities and states that have received grant assistance from foundations and government agencies. Throughout the Fall of 2012, I have worked with groups in places as diverse as Utica, New York, New York City, Yolo County (Davis), California, Las Vegas, Nevada, and Tennessee to build strong foundations for new food policy councils. Prior to this fall and under the auspices of CFSC, I served dozens of communities by giving keynotes, conducting on-site workshops, participating in webinars, and consulting through email and phone calls. These services were directed at a variety of places from statewide initiatives in California, Georgia, and Wisconsin to communities like Dallas, New Haven, and Omaha.

Looking at the outputs for 2012, we can see the number of people and unique communities participating in FPC capacity building events:

· January to August (through CFSC and Mark Winne Associates): 2,097 participants in 28 communities
· September to December (Mark Winne Associates): 1,485 participants in 17 communities
· Totals: 3,582 people in 45 communities
As evidence that the food policy council movement is now international, the above numbers include consulting trips to South Korea, Australia, and Canada where food policy council initiatives are underway and, interestingly, receiving more government support and interest than they do in the United States.
In addition, the following achievements are also noteworthy:
· There are currently 503 food policy council listserv subscribers (to subscribe, contact Mark Winne at win5m@aol.com)
· Published the "how-to" FPC manual Doing Food Policy Councils Rights: A Guide to Development and Operation
· Co-produced with the Harvard Food Policy and Law Clinic the Good Laws, Good Food policy guides, one on local food policy and one on state food policy

The above guides are available for free download at www.markwinne.com.

Looking ahead, I expect to see a continued increase in the number of food policy councils as well as growing demand for capacity building assistance, both for development services as well as improving their response to ever expanding policy opportunities. Due to more citizen interest in local and state food policy issues as well as the recognition that just and sustainable food systems don't happen without intentional action by activists and policy makers, food policy will find itself occupying more space on public policy agendas.

Finally, I am hopeful that a new partnership between Mark Winne Associates and a major institution will lead to a substantial expansion in the ability to serve both the existing and emerging food policy council movement. Until that time, Mark Winne Associates is happy to assist individuals and organizations with their food policy council needs.

Mark Winne
www.markwinne.com
Author of Food Rebels, Guerrilla Gardeners, and Smart-Cookin' Mamas: Fighting Back in an Age of Industrial Agriculture(Beacon Press 2010)

BY EMILY TAREILA, SF, CA 2012

ET

An era of debt
Rides in on goat's backs
Is swapped for a sea shell
In the new grazing field

–Mariette Lamson

MAY

———

THUNDER

MAY 2013

zodiac signs

Cancer	Good for plant-ing, irrigating, grafting & transplanting	Scorpio		Pisces			Good for root crops but a very nice time to make nourishen
Most Fruitful			Fruitful		Fruitful		
Taurus	A time to plant lettuce, cabbage, sturdy stalks & root crops	Capricorn	Good time to plant tubers & root crops	Libra			Time for seeding of hay, grain, flowers while moon's beauty is crescent
Semi-fruitful			Semi-fruitful		Semi-fruitful		
Aries	Time for cultivating, pruning where weed destruction is sought after. Chickens hatched now will be active	Sagittarius	Best for destroying unwanted plant life	Aquarius			Suitable for seeding / destroying of unwanted growth
Semi-barren			Barren		Barren		
Virgo	Good for bringing order to the garden; general cleanup	Gemini	Good to cultivate veg / destroying weeds & growth	Leo		A time for killing weeds; unwanted growth. Cider should be drawn off for storage · A good time to cut hay · If curly hair is desired, cut hair in the increase of the moon.	

8

♅ ♈ W
♈ ☉ 5:46 / 20:00

---

1 — A solar eclipse occurs today, though it will not be visible from the Northern Hemisphere. Avoid sowing seeds today after 8 am through tomorrow.

♐ ♑ W
♈ ☉ 5:54 / 19:52

9

NEW

♈ TH
♈ ☉ 5:55 / 20:01

16

♋ TH
♉ ☉ 5:38 / 20:08

17

♋ ♌ F
♉ ☉ 5:37 / 20:08

24

♏ F
♉ ☉ 5:31 / 20:15

---

25 — Look for a penumbral (partly shaded) lunar eclipse after midnight tonight. Avoid sowing seeds after midday today and all of tomorrow. PG

FULL

♏ SA
♉ ☉ 5:31 / 20:16

2

♑ TH
♈ ☉ 5:53 / 19:53

10

♈ ♉ F
♈ ☉ 5:44 / 20:02

18

♌ SA
♉ ☉ 5:36 / 20:09

26

♏ ♐ SU
♉ ☉ 5:31 / 20:16

3

♑ ♒ F
♈ ☉ 5:52 / 19:54

11

♉ SA
♈ ☉ 5:43 / 20:03

---

19 — Easterly winds from May 19 to the 21 indicate a dry summer.

♌ SU
♉ ☉ 5:35 / 20:10

---

27 — Root crops are best sown as the moon wanes for better root establishment.

♐ M
♉ ☉ 5:30 / 20:17

4

♒ ♓ SA
♈ ☉ 5:51 / 19:55

12

♉ SU
♈ ☉ 5:42 / 20:04

20

♌ ♍ M
♉ ☉ 5:34 / 20:11

28

♐ ♑ TU
♉ ☉ 5:29 / 20:18

5

♓ SU
♈ ☉ 5:49 / 19:57

---

13 — 1794: Whiskey Rebellion begins in western Pennsylvania. The War of Independence only just over, people who have recently become "Americans" take up arms against their own 'revolutionary' government & fight for autonomy. AG

♊ M
♉ ☉ 5:41 / 20:05

21

♍ TU
♉ ☉ 5:34 / 20:12

29

♑ W
♉ ☉ 5:28 / 20:19

6

♓ M
♈ ☉ 5:48 / 19:58

14

♊ TU
♉ ☉ 5:40 / 20:06

---

22 — As the full moon encourages germination, the sowing of seeds for above ground crops 2-3 days before the full moon provides optimal conditions for success.

♍ W
♉ ☉ 5:23 / 20:13

30

♑ ♒ TH
♉ ☉ 5:28 / 20:20

7

♓ TU
♈ ☉ 5:47 / 19:59

---

15 — In 1862, Abraham Lincoln signed into law an act of Congress establishing "at the seat of Government of the United States a Department of Agriculture." The establishment of the Department was the result of a long series of changes and improvements in American farming.

♊ ♋ W
♉ ☉ 5:39 / 20:07

23

♍ ♎ TH
♉ ☉ 5:32 / 20:14

---

31 — 1578: The Catacombs of Rome are discovered by accident.

♒ F
♉ ☉ 5:28 / 20:20

THE NEW OLD TRADITIONS
THROUGHOUT THE YEAR

MAY DAY: "The Joy of Beauty"
(May 1st)

May Day is the New Old Tradition of celebrating full-blown Spring. As our story progresses
we see the Sun grown into adolescence. Adoles-
cents being the most "wonderful" of creatures
to be around, The Sky and Earth act out in a
symphony of emotions. The Sky dumps buckets
of rain while the Earth shoots flowers from its
soil. The Sky gives reflection to the Sun's bril-
liance with a handful of glorious cool sunny
days, while the Earth offers her green grass to
whomever can find a plot. May Day is a day of
revelry.

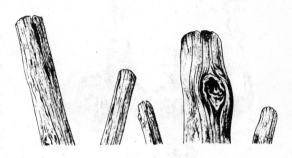

"Hut Happiness is a protective state sought
continually by some and occasionally by all as
an escape from events and prospects which dismay. The human being is most sufficient when
least in contact with conditions which dwarf
his personality and suggest the insignificance
of his tenure, his works and his emotions...."

-Clifford and John's Almanack

EM

Second Treatise on Civil Government (1689)

God hath given the world to men in common...
Yet every man has a property in his own person. The labour of his body and the work of his
hands we may say are properly his. Whatsoever,
then, he removes out of the state that nature
hath provided and left it in, he hath mixed his
labour with, and joined to it something that is
his own, and thereby makes it his property.

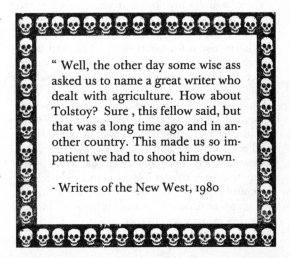

" Well, the other day some wise ass
asked us to name a great writer who
dealt with agriculture. How about
Tolstoy? Sure , this fellow said, but
that was a long time ago and in another country. This made us so im-
patient we had to shoot him down.

- Writers of the New West, 1980

Copper plate prints borrowed and printed from Rogersons Hardware Store, Hudson, NY

If Hairy Vetch Took Over The Farm Bill

by JANNA BERGER

The dirt under my fingernails may remind you of Laura Ingalls Wilder, but I am a twenty-first century farmer. I am part of a movement with a long-term vision for nourishing all human bodies, respecting the hands that grow our sustenance, and maintaining the earth as a hospitable, fertile home.

I may have looked quaint in my straw hat when I harvested them, but these veggies on the farmer's market table are just as much a part of the food economy as a box of Rice Crispies on a supermarket shelf. The hairy vetch that I just seeded among our cabbage plants is just as modern a technology in its response to the agricultural challenges that face us as the insertion of bacterial genes into a soybean seed.

Contemporary American agriculture is dominated by commodity monoculture, chemicals, genetic engineering, and synthetic fertilizers. This is not due to some kind of historical inevitability, but rather thanks to the power of the agribusinesses that benefit from high-input farming and the billions of U.S. federal dollars in research funding and direct payments that support and incentivize it.

Agribusiness lobbyists have a lot to do with who gets federal dollars, but so does the popular perception that industrial agriculture is more efficient and can best meet the food needs of our globe. Did the chicken come first, or the egg? Industrial farming receives overwhelming annual support and it feeds the world. What if a different kind of agriculture were the golden child of the farm bill instead? Even as an unconventional, renegade, colorful dot on the landscape of commodity monoculture, our small, diverse farm manages to grow thousands of pounds of fresh food each year. I can only imagine what we could do if billions of federal dollars backed our style of farming.

Funding for sustainable agriculture and "specialty crops" (which is government speak for fresh fruits and vegetables - the same crops that the USDA recommends as a basis for a healthy diet) is a drop in the farm bill bucket, but it does exist. How can we make modern ecological farming and the equitable distribution of fresh food central to the farm bill? How do we replace commodity subsidies and food stamps that help people just barely squeak by in food deserts with a comprehensive switch to a food system that serves more people better?

I hope to start a revolution by explaining the kind of leguminous nitrogen fixation that our hairy vetch accomplishes every chance I get, so here it goes. Nitrogen is an important nutrient that all plants need to grow. It is not always readily available to plants in the soil, so most farmers in America apply manufactured fertilizers that require massive quantities of fossil fuel to produce. These products provide a direct shot of nitrogen to plants. Because it is so soluble, whatever the plant doesn't use runs into our aquifers and watersheds causing algal blooms and anaerobic dead zones.

Through the symbiotic relationship they have with rhizobia bacteria, legumes like hairy vetch can grab nitrogen out of the atmosphere with their leaves and fix it into the soil. Along with the organic matter they produce, legume cover crops leave behind nitrogen rich, fertile soils that can feed cash crops over time.

It is early September and we just finished a mad hairy vetch planting spree, largely under the sage guidance of a book called "Managing Cover Crops Profitably." This gem of a tool was researched and published by the scantily funded federal Sustainable Agriculture Research and Education program. It is an eminent teacher in how to replace herbicides, pesticides, soil erosion and expensive fertilizers with soil building grains and legumes.

What if the government spent a little less money on military escorts for oil tankers that haul fossil fuels across the world to produce manufactured fertilizers, and the farm bill included a little more money distributing "Managing Cover Crops Profitably?" What if the government took on the mission of teaching farmers to grow "specialty crops" with the help of legumes while rewarding those who did so with conservation payments? What if the government helped farmers become more efficient with these systems and flood the market with cabbage and tomatoes and apples, driving down prices and making those life-giving fresh foods more affordable? What if billions of health care dollars were saved from diabetes and heart disease treatment because we focused our agricultural dollars on producing fresh fruits and vegetables?

But hairy vetch and direct support for specialty crop producers won't fix our food system on their own. We need the farm bill to incentivize consumer and institutional purchases of fresh foods - especially for those who can't otherwise afford them by bolstering and improving programs like the SNAP Fresh Incentive Program and the Fresh Fruits and Vegetables Program. We should support farmer-to-consumer direct relationships with programs like the Farmers Market Promotion Program as well as the viability of small, diverse farms by expanding the reach of programs like the Value Added Producer Grants and Rural Microenterprise Assistance Program. We should support school and community gardens in cities, and teach people to grow their own food. We need to disincentivize nitrate runoff, carbon emissions and other agricultural pollutants by passing on the expense of environmental cleanup to polluters rather than taxpayers. We should help disadvantaged farmers thrive and fund Beginner Farmer Rancher Grants, considering that the average age of the American farmer is fifty-seven. We should safeguard farm workers' rights and treat them with the respect that folks who nurture our sustenance into being deserve. We need to further develop public research and extension for sustainable technologies that benefit citizens rather than corporations, investing in beneficial insect habitat, crop rotation, composting and erosion control. We need to stop dumping excess grain on poor countries, running local farmers out of business and jeopardizing long-term food security for those populations. Rather, we should use our international influence to support vibrant regional food systems all over the world.

It sounds like a tall order, but we are talking about a piece of legislation that already shapes the way that people grow and eat food. Boiled down, that means that it shapes who we are, what is inside us and what sustains us. We should expect it to do so with love, care and mindfulness for us and for the earth under our feet.

I have heard the heavy question posed over and over again, "can organic, small-scale agriculture feed the world?" Under the current system one in six Americans and nearly one billion people on earth suffer from chronic hunger. One third of U.S. adults and nearly 17% of children suffer from obesity. So, isn't it worth imagining alternatives and testing the waters of what it would look like to shift paths?

Sustaining human life on earth is not simple. We are enormous, complicated animals who have a tendency to take up a lot of space. It would be hard for us, and our big, busy minds, to feed our intricate society on instinct alone, without learned knowledge. We are not irreconcilably compelled to ram our faces into tree trunks repeatedly, digging for insects like a woodpecker, nor are we innately motivated to bury acorns underground for winter safekeeping like a squirrel. We are not born out of eggs, laid scrupulously by our mothers on a delicious buffet of our favorite food like a tomato hornworm.

Rather, human food production and distribution is built upon generations of innovation that must be taught. I don't suggest that we go backward in time, to try to farm the way our grandparents did because we've gained so much wisdom since. America bears the history of massive land-theft and slave bondage in the name of agricultural production. With such a heritage, Americans are all the more called on to be judicious students, comparing the implications of different systems on the tangled web of socio-economic and environmental concerns. A growing movement of farmers and consumers is taking a compassionate, long-term, connection-based approach: combining our accumulated scientific and indigenous knowledge in conscientious ways and searching out the best farming models for our time. It is time to get the weight of the farm bill behind us.

See.

One.

Grow.

Delight.

Build.

Birth.

Know.

Arise.

Love.

CGK

FIXED FACTS IN AGRICULTURE.

The following list of "fixed facts" in agriculture, for once, in a condensation of the sort, hits the right nail on the head in most of them:

1. All lands on which clover or the grasses are grown must either have lime in them naturally, or that mineral must be artificially supplied. It matters but little whether it be supplied in the form of stone-lime, oyster-lime, or marl.

2. All permanent improvement of lands must look to lime as its basis.

3. No lands can be preserved in a high state of fertility, unless clover and the grasses are cultivated in the course of rotation.

4. Mould is indispensable in every soil, and a healthy supply can alone be preserved through the cultivation of clover and the grasses, the turning in of green crops, or by the application of composts rich in the elements of mould.

5. All highly concentrated animal manures are increased in value, and their benefits prolonged, by admixture with plaster, salt, or with pulverized charcoal.

6. Deep ploughing greatly improves the productive powers of every variety of soil that is not wet.

7. Subsoiling sound land—that is, land that is not wet—is also eminently conducive to increased productions.

8. All wet land should be drained.

9. All grain crops should be harvested before the grain is fully ripe.

10. Clover, as well as the grasses, intended for hay, should be mowed when in bloom.

11. Sandy lands can be most effectually improved by clay. When such lands require liming or marling, the lime or marl is most beneficially applied when made into composts with clay. In slacking lime, salt brine is better than water.

12. The chopping or grinding of grain, to be fed to stock, operates as a saving of at least twenty-five per cent.

13. Draining of wet lands and marshes adds to their value, by making them to produce more. and by improving the health of neighborhoods.

14. To manure or lime wet lands, is to throw manure, lime, and labor away.

15. Shallow ploughing operates to impoverish the soil, while it decreases production.

16. By stabling and shedding stock through the winter, a saving of one-fourth the food may be effected : that is, one-fourth less food will answer than when the stock may be exposed to the inclemencies of the weather.

17. A bushel of plaster per acre, sown broadcast over clover, will add one hundred per cent. to its product.

18. Periodical applications of ashes tend to keep up the integrity of soils, by supplying most, if not all, of the organic substances.

19. Thorough preparation of land is absolutely necessary to the successful and luxuriant growth of crops.

20. Abundant crops cannot be grown for a succession of years, unless care be taken to provide an equivalent for the substances carried off the land in the products grown thereon.

21. Young stock should be moderately fed with grain in winter, and receive generous supplies of long provenders, it being essential to keep them in a fair condition, in order that the formation of muscle, bones, etc., may be encouraged and continuously carried on.

HOW TO BARTER *by* CHRISTIE YOUNG

MAKE IT... TRADE

4. MAKE AN AGREEMENT.
WHO / WHAT / WHEN - YOU
HAVE NEEDS, THEY HAVE
NEEDS, FIGURE OUT HOW
TO SOLVE THEM & SHAKE
ON IT.

1. DO YOUR RESEARCH.
WHAT'S THEIR SIGN?
RIGHTIE OR LEFTIE,
INNIE OR OUTIE?
KNOW BEFORE YOU
GO...

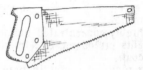

5. JUST DO IT.
NO ONE LIKES A MOOCH,
A BLOWHARD, OR A CHEAT.
DO WHAT YOU PROMISED!
YOU SHOOK ON IT,
REMEMBER?

GREAT JOB!

2. MAKE CONTACT.
MENTAL, EMOTIONAL,
A LIGHT TOUCH ON
THE ARM... START A
CONVO, IDEALLY
WITH A WINK (OR
VIABLE BARTER)

6. LEAVE FEEDBACK.
ONLY ONCE YOU & YOUR
PARTNER (B-FRIEND) AGREE
IT'S COMPLETE (OVER).
BE HONEST, BE FAIR, &
BE YOURSELF.

3. TALK IT OUT.
DEADLINES, EXPERIENCE,
WHAT'S CONSIDERED
"FAIR" IN THE RELATIONSHIP.
REFER TO THEM AS
YOUR "B-FRIEND"

And I will send grass in they fields for thy cattle
that thou mayest eat and be full.

DEUTERONOMY 11:15

How Farmers Can Cooperate

FROM FARM ENCYCLOPEDIA 1889

Successful cooperation depends upon the mental attitude of the people who are to become members of the organization. Just as some men are tall and some are short, so it may be said that some are naturally good cooperators and others are not. Each member must be not only willing, but anxious, to give and take. Selfishness must be overcome. The desire to get the best of a neighbor must be eliminated. The motto must, to a very large extent, be, "Each for all and all for each." The greatest success for the organization will be measured by the extent to which the members willingly adopt this motto and actively live up to it. Greater success will come from 10 farmers who have the inclination to cooperate and actively live up to their faith than from an organization of 20 farmers, each suspicious of the other's good intentions and willing or anxious to take advantage of the other if an opportunity presents itself.

In any community the meeting of the minds of the people must be the starting point of cooperative organization. All must agree as to which is the most important, most practicable, and most useful form of cooperation for the community. After this question has been settled, and after a sufficient number of neighbors have become sufficiently acquainted to be sure that the cooperative mind or attitude is present, the actual organizing – the drawing up and adopting of articles of incorporation, constitution, and by-laws, and, finally, of rules and regulations for the government or management of the enterprise – all this is an easy matter. A letter to any college of agriculture, or to the Department of Agriculture at Washington, D. C. will bring sample copies of forms which may be used almost without change. Many collections have been made of forms which have been successfully used in hundreds of communities, and these need not be reproduced here. Several good books on cooperation among farmers are available, some of which carry all necessary forms, blanks, sample constitutions, etc.

Organization should not be rushed. It is more important to devote weeks, even months, or a year or two, to the task of organizing and perfecting plans than to try to rush through these plans hastily and imperfectly at one brief meeting, in order to engage quickly in some form of cooperative enterprise. Probably more failures are due to haste in preparation than to any other cause. Too often a group of neighbors, called together on the spur of the moment, adopt something which is read, and immediately wish to see results. Unless they secure very large returns within the first few months, some are disappointed and fall by the wayside, others become discouraged and commence to complain. From years of experience, study, and observation, it is now positively established in my mind that patience, forbearance, and per- severance – all of which mean willingness to go slow and to lay a solid, permanna, and thorough foundation–are the fundamental necessities, if great results are to be secured in the long run.

Cooperation in farming for business purposes shows signs of 2 conflicting movements. Some forms of cooperation have for their aim the bringing back to the farmers some of the work which they formerly had and which was taken away from them by the development and centralization of industrial enterprise; other forms aim to relieve the individual farm or farmer of tasks which he can clearly not perform most successfully or most economically. In the case of cooperative elevators, for instance, success is based upon the fact that individual farmers cannot most successfully or economically store their products on their own farms preliminary to placing them upon the market.

At the same time, it is based upon the confident belief that farmers can and should retain control and ownership of their products until the best time for marketing arrives. In other words, it is based upon the established belief that a better market can be secured and better prices received, if farmers do not immediately sell the products on their farms to outside parties, but rather that, if they will hold their products until the market is ready, more satisfactory results will be secured. The success of the cooperative creamery, cheese factory, cannery, etc. is based upon the well-established principle that the manufactured product —butter, cheese, canned vegetables, etc. — will be more uniform, better prepared, if made in a small local factory than it will be if made on individual farms or even in great central establishments.

The whole success of the great new movement of cooperation in farming will be based upon successful analysis of the best form of organization in each case. In the matter of flour milling, meat packing, cloth making, etc., it will probably prove true in the long run that large central establishments are more economical and more efficient than small local establishments close to the farm. The larger central establishments will probably produce more cheaply and produce a better article and at the same time bring the finished product to the consumers at a lower price. In the matter of making butter, cheese, canned goods, etc. on the other hand, the local establishment, locally formed and operated, will probably be the more successful. This is all the more reason why time should be taken in each case in definitely determining the best program for each community.

A RESOURCE FOR COOPERATIVE STARTUP & RESEARCH
WWW.CULTIVATE.COOP
USDA COOPERATIVE DEVELOPMENT GRANT PROGRAM

EDITORIAL NOTE: This encyclopedia inspired the OPEN Farm Data wiki project, begun by Dorn, Severine, Henry, RJ and a few others in 2012. Basically we are working on collecting, interpreting, connecting and re-using historical farm advice by putting out of copyright books into a wiki so that we may resume a "relic trajectory" of progress and good insight at a moment in time before chemical agriculture. Most of what they wrote in this Encyclopedia is still true and in a wiki we can add in what has been discovered lately. To join the ag history ratpack, please be in touch with farmer@ thegreenhorns.net, and bring your wildest ideas. -SvTF

INGALLS

"The story goes that an old Nebraska farmer was sitting on his porch during a dust storm. Asked what he was watching so intently, he replied: " I'm counting the Kansas farms as they go by."

- STUART CHASE MAXIMILLIAN 1927, *from Rich Land Poor Land*

AML

"Agricultural land is a trust inherited by those who possess it today, to be used while they live, and to be bequeathed in at least the same, if not better, condition to those who follow then. The idea that the present -day possessors of land hold it absolutely, and that it is there to monopolize its value and exhaust its fertility, is not only ethically and esthetically false, but economically unsound. Each generation of farmers consists of tenants for life only. It is impossible for any generation to speculate in land or to exhaust the soil entrusted to it without depriving its own children and all future inheritors of the earth of their birthrights. Yet having accepted the idea that both land and farming are primarily methods of making money, that is precisely what we are doing, and what we are today being urged to do in an even greater mesa sure by most of the agricultural leaders who are telling us what is essential in order to modernize agriculture."

- RALPH BORSODI *from Agriculture In Modern Life*

One Farmer's P.O.V. The Federalization Of Organic

You may remember when we asked for your help in 1998 to stop the USDA from undermining organic labelling with its "National Organic Program" (NOP). A quarter million people who participated in that public comment process kept the NOP from allowing things like genetically engineered plants to be marketed as "organic".

The revised program is finally coming into effect as of next Spring, with standards that legally supersede the many local definitions of "certified organic" that had been created at the grassroots level over the past 25 years. In their current form, these standards are deeply flawed, if not obviously absurd. Indeed, the fact that they are not a totally obvious corruption of everything tha "organic" has previously stood for may make them that much more threatening to our movement.

Since 1990, Chris has regarded the NOP as an effort by big business interests to steal a profitable (and increasingly threatening) "brand identity" from the hundreds of thousands of consumers, farmers, and activists who actually did the work to "buil the brand". From this point of view, it does not make sense to participate in the program even in the unlikely event that the USD/ writes an acceptable set of standards.

Where did this mess come from? Back in the late 80's, there was an official line from the private lobbying group that put the pressure on for the federal program (a group called OFPANA, now re-named OTA - the organic trade association): they claimed that without nationally homogenized organic standards, the organic label would be dragged down to meaninglessness (like "natural" in the '70s) by fraud and commercial pressures. At the time many of us saw no evidence of wide-scale fraud, nor did it appear inevitable that local private certifying agencies (like NOFA MA) would fail to uphold high organic standards. It turned out that, after the NOP law was passed in 1990, the grassroots certifiers did a fine job of protecting the integrity of organic labelling. They did great for 10-years of double digit organic growth while the USDA wrote its regulations. If we don't believe the lobbyists' official line, what might have been the motives in asking for federal homogenization of "organic"? Of course we can only speculate But a few observations about the marketplace before and after can offer some insight. Organic certification based on locally controlled standards was an obstacle to parties who envisioned "organic" enterprises on a national (and international) scale. It wasn't an impossible obstacle. Plenty of products sell in MA with certification by, say, Oregon Tilth. But Massachussetts consumers, faced with an array of certification seals, will likely trust the local organization before others from further away (if they bother to think of certification at all). Because of this, companies with their eyes on mass markets (mostly not farmers, but processors and handlers, and probably crucially, their prospective creditors) may have judged the old organic marketplace to be insufficiently predictably profitable. If so, the national standards that we actually ended up with are certainly consistent with an effort to "make th world safe for 'organic' agribiz".

How so? In the first place, they take the power to decide the rules by which we will participate in an alternative (previously known as "organic") marketplace out of the hands of decentralized groups somewhat accountable to popular influence (fc example, NOFA MA), and put that power into the hands of a centralized beaurocracy (USDA) that is in turn appointed by the executive branch, which, although technically "elected", is in practice extremely well insulated from popular accountability (I'm not referring here to extraordinary events like the 2000 election, but to the undemocratic features of the average national election, one of the most obvious being the "money primary"). It's also interesting to observe a familiar pattern in this process, namely that the us of state power by big business to enhance their prospects for profit invariably entails concessions on individual rights for the rest of us. Thus, our cedeing control over the word "organic" is substantively equivalent to giving up rights of free speech and free association in this matter. Meanwhile, for example, the tobacco companies reserve the right to free speech on behalf of their lethal products.

Furthermore, the production details of the new rules also happen to discriminate against small producers in many ways, consistent with the hypothesis that they were written mainly as a mechanism to enhance the profits of national-scale companies. One of the areas where the new rules will have a big impact on the Northeast is composting. Even though you helped us keep sewage sludge off organic farms, the new rules treat animal manure as though it were sewage sludge, and use of manure is subject tc regulations apparently lifted from EPA guidelines for municipal sewage composting. These regulations impose criteria (e.g. maintenace of pile temperatures at 140 for 5 days) that no one acheives in practice without equipment investments of tens, if not hundreds, of thousands of dollars. Without such equipment, small growers must handle any compost containing some portion of manure as though it were 100 percent uncomposted ("raw") manure, which is subjected (appropriately) to its own regulations. The principal rule (which NOFA already had) for raw manure is that there must be an interval between application and harvest of 120 days for crops whose edible portions are in contact with the soil (90 days for crops not touching soil). This means many growers will be applying manure the fall before planting. While it would be technically allowable to apply this manure uncomposted ("on an actively growing cover crop"), conscientious (and practical) growers will be composting the manure before a Fall application, because we know that we would inevitably lose soluble nutrients from Fall manure application, which of course become pollutants.

This means the composting process will have to take place during the prior growing season, using up the small grower's valuable time and land whereas before one could simply compost over the winter prior to spreading in the Spring.

An ironic footnote to this situation is that it is likely that the USDA would have faced more criticism about its decisions on composting rules if it weren't for *e. coli*.. The small farmers' spokespeople on the National Organic Standards Board (an advisory board) probably kept quiet about this issue out of fear of bad publicity associating organic farming with manure and *e. coli*.. This fear was in part due to misinformation campaigns by industry funded "think" tanks, a recent particularly egregious example being a 20/20 TV episode which the network later had to retract.

The upshot of all this is that Chris will likely <u>not</u> continue to certify his side of the operation, while Tom will likely continue when he returns to farming because his wholesale markets must be able to legally label his product as "organic". We will both continue to farm in much the same way we have always farmed, with subtle differences (such as in composting techniques) arising in areas where the new standards depart from the old. You have our assurance that all produce coming from each of our operations will still meet or exceed the old NOFA MA standards for certifed organic. (Legal note: this last sentence may technically be a violation of federal law. As far as we know, it is still legal for you, our customers, to refer to our vegetables as "organic".)

We feel our farm's response reflects a few important truths that will be helpful to keep in mind in the course of the ongoing struggle to build an alternative food system for our region. First of all, it is of course a good result that more and more production on faraway large scale farms will occur in a more ecological fashion. The "organic" label isn't useless yet. But it's now even more clearly insufficient for acheiving the movement's historic goals of creating an alternative food system compatible with elementary notions of ecological sustainability and social justice. To be sure, there were many people, (such as the family Chris apprenticed with way back in 1988), who reminded us that this was the case even as the first formal <u>local</u> certification programs were being born. Now the implementation of the NOP is a golden opportunity to heed their call to take the organic concept "to the next level". The problem is that it's slow, hard work, often prohibitively so in the context of small farms which need to make money right away and don't have much labor to spare for movement building. Somehow, we succeeded in spite of these obstacles in the case of the "organic" concept, and the challenge for a lot of us now is to take the next step. What form will this next step take? Will it be new labelling schemes promoting "locally grown", "sustainably grown", "fair trade", "authentic", "union made"? Will it be a proliferation of non-profit advocates like "CISA" (which runs the "Be a Local Hero Buy Local Produce" campaign in Western MA), or NYC's "Just Food" (which connects city people with upstate CSAs), or Quebec's "Equiterre"? Will it be a resurgence of local independent store fronts (like our own Harvest Co-op or Portland's Public Market) where consumers will be able to find the alternatives to mass-produced 'organic' that they demand (and maybe pick up food free of GMOs but not necessarily organically produced)? All of these innovations will probably play a role, and no doubt some that we can't even imagine (anarchic internet clearinghouses with unlimited information on production practices?).

At the same time, it would be tactically foolish for small growers to abandon the toehold they have in mass markets by completely renouncing certification. Every dollar that is shifted from the corporate sector to the small producer is vital. Besides the dollars, the movement would also be losing the potential to connect with a wider audience of customers who may as yet not differentiate between organic products that come from local family farms and products that come from far off industrial scale operations controlled by corporate interests. So our movement will include many farmers such as Tom who fight a rearguard action in the mass marketplace, trying to defend the niche for small growers that exists.

For our farm, our CSA system is the single innovation that gives the most promise for future viability. We expect CSAs to continue to proliferate and play an important role in our movement's response to this situation. Not all farmers can have or want to have CSAs, so we will need the other creative solutions as well. All of these solutions will likely share some element of increased customer education. I imagine many customers who buy an organic product intend to support a family farm with their purchase. We need a way to distinguish our produce from "industrial organic" to serve that latent demand, and to build awareness of the issues involved so the demand grows. To quote our 1998 letter "the best defense ... is for consumers to become more knowledgeable about the facts of food production".

Federalization Is Fuedalization

by DAVID STERN

Rose Valley Farm, Spring, 2002,
From the Publication of the Northeast Organic Farming Association

I started farming in 1970 and working on standards for NYS organic certification in 1984. I've spent hundreds of hours at my desk reading and studying. I've spent hundreds of hours around kitchen tables in dialogue and formulation. I've spent hundreds of hours on the road. All unpaid. These were exciting times of high energy when a small group of farmers and friends would debate for hours on the great questions of the day: black plastic, percent organic grain for livestock, hydrogen peroxide, transitions and the 36 month rule, raw manure, sulfites, and costs. The program reached out to all who wanted to participate. We tried to capture an ethic and spirit, - to put it on paper, in black and white: applying organic principles to an administrative program. Looking for the balance between production and the consumer...and reality. Robert Perry recently referred to us as a "some-what intractable group of farmer-philosophers" Yup. After reviewing domestic and international programs we stole what we liked, we changed what we needed to, and we filled in the gaps to hold it together. This was a process not all New York organic farmers supported, but we listened and tried to meet as many concerns as we could. The NOFA program grew from seeds planted by the farmer, not the consumer. It was not easy. Other states had Government/Ag and Markets support. We had none. Other states had Land Grant/Ag School support. We had none. What we did, we did on our own. We have been blessed with a skilled administrator and incredibly dedicated volunteer board members who took the pages of black and white and made it real. We can all be proud of our labors. We gave life to a program and breath to a very small industry. I share this introduction with you as a way of saying that I've paid my dues (and certification fees for 15 years), and now feel the time has come to express my opinions as a matter of obligation to this industry, and to encourage debate and comment. I do not speak for the others who sat at the table in this or other states and programs, or for my cooperative. I speak from the tradition of independent rural populism and I speak from my heart. I am tired of listening to people who preach "local/regional food systems", "small is beautiful", and "sustainable communities" on Sunday morning and who are on the phone to the USDA on Monday afternoon. Maybe it is time for a new perspective to this issue: that instead of capitulation and submission to the USDA, we rise from our knees, we stand up, and we say "NO" to this federalization which has become feudalization; "No" to the USDA agricrats who have bastardized the organic principles and sucked away the energy from local decision making; and "No" we will no longer cooperate. We always have a choice in all that we do. The only time you can't change direction is when you're dead. The Feds have stolen the word "organic" and unless we play by their rules, we can't use it. O.K. How about "organik", "cinagro", "orgreen", or better yet, "orgasmic"! They have stolen a word; let's not give them our spirit. Let's not be distracted by the nonissues of irradiation and sewage sludge. Let's be thankful for the financial help from Ag and Markets with the understanding it's only to lessen the shock of increased fees. Look through the smoke at the realities that we are facing. Let's support the 5% of 10% of us who sell to processors on interstate/international markets, but not to the detriment of the entire industry. Let's educate the consumers who support us, but don't understand our reality, that this feudalization is not in the best interest of the organic farming community and will only further depress our already fragile economy. Let's not allow our strengths of "independence, stubbornness and fortitude" remain our weaknesses. Some of us have never been in support of the federal legislation and understood when the USDA appointed the very first National Organic Standards Board that a real monster had been created. Initiated in 1990 by our brothers and sisters who work those beautiful Vermont farms, it was a monster we could not control or even influence. The snowball of bureaucracy was rolling downhill through the corridors of the USDA, soon to be an avalanche upon us. We hadn't the time, money, influence or lobbyists to do anything but watch and wait. We are only small hardworking farmers wanting to carve out a small space in the marketplace. Some of us don't have computers or ease of words to express our thoughts, or even have clear organized thoughts on the 500 pages of regulation we never read! Some of us need to concentrate on increased productivity, better land stewardship, CSA's, processing, cooperatives, making a living at work off the farm, community work, and even time with our families. We will never read those 500 pages. However, the legal staffs of the Mega-Capital/Industrial/Agricultural Complex read them and watched as that shelf space grew 3…4…5 percent. They used their lobbyists and influence to formulate the rules and procedures and their money to buy farms, farmers and cheap labor. The big fish will eat the little fish. In the great spirit of Sam Walton, I refer you all to "Between The Furrows", The Natural Farmer, Spring, 2001. The USDA National Organic Program is a long, long way from those kitchen tables. As Jim Hightower said, "The water won't run clean 'till we get the pigs out of the creek!" We will soon hear about the number of farms and number of acres in the Federal program (as if the USDA should take credit). These are the values that are important to the Mega-Capital/Industrial/Agricultural Complex. I propose we look at the farms who drop away from certification, who stop their organic transition, or who will never consider it an option for their operation; let these be the "measure" of the success of feudalization. It's been over ten years ago now since the members of the NOFA Organic Standard Board walked to the upper fields at Hemlock Grove Farm in West Danby and talked

about the "worst possible scenario". The USDA hadn't put any money into the National Organic Program (and didn't want to), so our fingers were still crossed. But we asked "How bad could it get?" and that's about what we got. We knew then that the issues of compost, fees, role of farmers, material approval, inspectors, seeds, state relations, decision-making, etc. would not be decided in the best interest of the small farmer. City folks have these events: "TAKE BACK THE STREETS", and "TAKE BACK THE NIGHT" when they stand together and face the evils of their communities. When those responsible have failed to act in their best interest and for the good of their neighbors, they say: "Enough! No more." I suggest we rise and stand together, as caretakers of the earth and those blessed to grow life-giving food, shed our fear of the monster and the unknown, and say, "NO!"

EDITORIAL NOTE: The opinion of the author does not indicate the position of the editor of this Almanac, but we include these two pieces as a reminder of the value of open debate within the organic sector, and the rights to free speech about production methods. Farmers are urged not to talk about production practices (by the Farm Bureau) in order not to disturb the consumer, not to criticize other producers methods, to present a unified front for the small minority of farmers in this country to speak with one voice. Not everyone shares this opinion, so we include in this volume these independent voices of farmers, having a legitimate rant. An almanac is an appropriate place to do that. - SvTF

COMMENTS ON ORGANIC STANDARDS

By Eliot Coleman

My opinion on this topic is the same today as it was when this national process began. There is a better way of achieving cleaner, more nutritious food for consumers than imposing a national definition of "organic."

This better way, letting individual labels define themselves, was the practice in Europe during the '70s and '80s. The various European "organic" organizations -- *Nature et Progres, Lemaire-Boucher, Soil Association, ANOG, Bio-Organic, Demeter* -- each defined and published standards to which their food was grown, based on their different theories of how to produce the best quality food. There was even a Swiss supermarket chain, Migros, with its own line of low-chemical-input foods called "Migro-sano." Migros contracted with Swiss farmers to grow food to specific standards which banned the chemical inputs Swiss consumers were most concerned about, while allowing the less toxic products.

This open system offered numerous advantages to European consumers. Not only was there a range in price and quality, there was also the power to continually upgrade the standards. Whenever new agricultural research raised flags about a previously acceptable input or practice, the consumer shift to the labels not using that input or practice forced the other labels to shape up. This was a system driven to become ever better in response to the concerns of astute consumers rather than, as with any politically controlled system, ever more watered down in response to the influence of the powerful lobbyists.

Consumers should be aware that the virtues of this successful European model are presently seen as its fatal flaws. Such a wide range of consumer opinions and flexibility for improvement is unacceptable now that "organic" is big business. The expanding "organic" industry needs one simple, lowest-common-denominator definition for international trade. That is what the new Organic Standards are. But my question is this: Shouldn't the "organic" food option place the benefit to consumers ahead of the needs of corporations?

TRANSPLANTING EVERGREENS.

It seems not very material whether evergreen trees are transplanted in April, May, or June. They may be made to live in either of these months, when they are properly taken up and set ; and as it is all-important to take up a sod with the tree, it may be as well to transplant this kind early in the season before plowing commences.

It is not necessary to take up a long root with a fir, a hemlock or a pine ; but it is absolutely necessary to take up a sod with the roots; and sods will adhere to them better at this season of the year, than when the earth is more dry.

There is not much risk in taking firs from good nurseries, for the multitude of fibrous roots that are found in every direction, hold enough earth to insure their growth. But pines or firs taken from forests have but very few roots, and they need more care.

The bark that covers the roots of pines and other evergreens, is very thin and tender, and when the trees are *pulled* up and set, as we set apple trees, the bark comes off, and not one tree in fifty survives. Long roots are not needed, and the trees may be taken up by cutting around at a distance of twelve inches from the trunk, when that is not more than five feet in height.

These trees and clumps of earth may be set when the earth is wet, for there is not the same need of spreading out the roots and keeping them separate, as there is when trees are taken up without earth. Yet it is important in all cases to keep the earth loose, and light, and free from weeds around them.

FORMS OF VEGETABLE GROWTH.

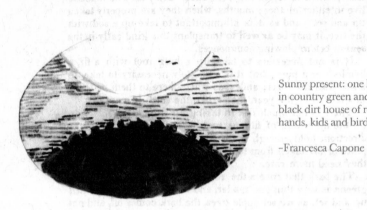

Sunny present: one long day
in country green and city
black dirt house of reaching
hands, kids and birds

-Francesca Capone

JUNE

—

SLEET

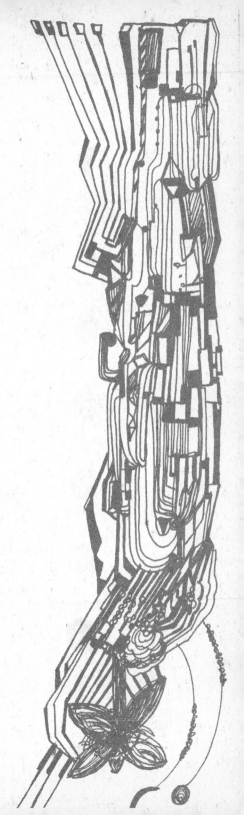

JUNE 2013

1 ♒ ♓ SA ♉ ☉ 5:27 20:21

2 ♓ SU ♉ ☉ 5:27 20:22

3 ♓ M ♉ ☉ 5:26 20:23

4 ♓ ♈ TU ♉ ☉ 5:26 20:24

5 ♈ W ♉ ☉ 5:26 20:25

6 ♈ ♉ TH ♉ ☉ 5:25 20:25

7 ♉ F ♉ ☉ 5:25 20:25

8 1997, Cambridge: Anti-genetic food activists play cricket using bioengineered potatoes previously sched-uled for harvesting. What with a muddy field & hard swings, the entire crop is destroyed. [NEW] ♉ SA ♉ ☉ 5:25 20:26

9 AG ♉ ♊ SU ♉ ☉ 5:25 20:26

10 ♊ M ♉ ☉ 5:25 20:27

11 ♊ TU ♉ ☉ 5:24 20:27

12 ♋ W ♉ ☉ 5:24 20:28

13 ♋ ♌ TH ♉ ☉ 5:24 20:28

14 On June 14, 1854, Townend Glover, an entomologist, was appointed "for collecting statistics and other infor-mation on seeds, fruits and insects." A contribu-tion by him appeared in the 1854 report. ♌ F ♉ ☉ 5:24 20:29

15 ♌ SA ♉ ☉ 5:24 20:29

16 New Moon and first quar-ter Moon are considered fertile and wet. ♌ ♍ SU ♉ ☉ 5:24 20:30

17 ♍ M ♉ ☉ 5:24 20:30

18 ♍ TU ♉ ☉ 5:25 20:30

19 ♍ ♎ W ♉ ☉ 5:25 20:30

20 ♎ TH ♉ ☉ 5:25 20:31

21 Summer Solstice
Midsummer Night, The longest day of the year, is endowed with great mystical powers. According to Icelandic folklore cows gain the power of speech for the night, and seals can take a human form. ♎ ♏ F ♊ ☉ 5:25 20:31

22 ♏ SA ♊ ☉ 5:25 20:31

23 PG
The full moon occurs today at perigee (when the moon is closest to the Earth in its orbit), and thus will loom large in the sky. [FULL] ♏ ♐ SU ♊ ☉ 5:26 20:31

24 1976 · Poland: Government hikes staple food prices enormously. The response was immediate: nationwide strikes, public protests & rioting. ♐ M ♊ ☉ 5:26 20:31

zodiac signs

Cancer ♋ Most Fruitful	**Scorpio** ♏ Exceptional for vine-type growth	**Pisces** ♓ Fruitful
Taurus ♉ Semi fruitful	**Capricorn** ♑ Semi fruitful	**Libra** ♎ Semi fruitful
Aries ♈ Semi barren	**Sagittarius** ♐ Semi barren	**Aquarius** ♒ Barren
Virgo ♍	**Gemini** ♊	**Leo** ♌

25 The time period 12 hours before and after a solar or lunar eclipse are unfavor-able for sowing seeds. A partial lunar eclipse occurs tonight, though it will not be visible in North America. ♐ ♑ TU ♊ ☉ 5:26 20:31

26 If there are many falling stars during a clear summer evening, expect thunder. If there are none, expect fine weather. ♑ W ♊ ☉ 5:27 20:31

27 ♒ TH ♊ ☉ 5:27 20:31

28 ♒ ♓ F ♊ ☉ 5:27 20:31

29 ♓ SA ♊ ☉ 5:27 20:31

30 ♓ SU ♊ ☉ 5:28 20:31

THE NEW OLD TRADITIONS
THROUGHOUT THE YEAR

GIVING

MIDSUMMER: "Moon's Conception"
(On/around June 21st)

Even with some of its best days still to come, at Midsummer the Sun has reached its peak, and thus marks the longest day of the year. Since its birth the Day had been courting the Sun with flowers, sweet rain, and cool breezy whispers. When the time had come for the Sun and Day to make the sweetest of love, the Sun, so taken with the Day's offerings, refused to allow the Night to arrive lest its affair with the Day end too quickly, and so stretched the Day far into the Evening. By sundown the Moon had been conceived. In honor of this event we celebrate the beauty of Earth's bounty. Merriment abounds.

There are three kinds of givers in the world: the flint, the sponge, and the spring.

The flint has worth inside, but it must be hammered to release its spark. But the benefit of the spark—the gift of fire—is precious. Some givers are hard, but the precious potential is always there.

The sponge must be squeezed to give up its contents. The more you squeeze, the more you get. But once squeezed dry, the sponge must renew itself before giving again. Some people, too, are good-natured sponges who yield readily to pressure. But they must have their own source of self-renewal.

The bubbling, sweet spring seems never to cease its giving. It is a constant delight to those who know it, nourishing all life freely and equally. Some people delight in giving, and of these, the Bible says, "The Lord loveth a cheerful giver."

Grey Fox

RECIPE - Summer

by ZOE LATTA

grate an english cucumber
put it on everything

"BUT. . ."

I am the "butter." I get my name by injecting the word "but" into what otherwise would be high compliments for people and things. "Yes, he's a fine young man; one of my best friends. I like him a lot, but . . ." "Sure, he's a fine preacher, and the church likes him, but . . ."

"Yes indeed! America is certainly all right, but . . ."

I'm a crack shot, I'm not going to stand by and let anybody or anything get away with unlimited approval! There are no closed seasons on compliments. I shoot them down anytime with my distinguished slander-gun. I never miss a shot!

—Western Recorder

farm games!

Some Games To Play
by MICHAEL BEGGS AND FRITZ HORSTMAN

ROCK WALL BUILDING
One on One

Playing field: 10 x 3 meters mostly flat cleared ground,
piles of stones at either end of the longer axis.
Equipment: 500 dark stones, 500 light stones in separate piles

The game: Each player begins next to his or her pile of stones. At the official's whistle the players begin building a wall. They are building towards each other, and will ultimately create one wall. The wall must be at least 50 cm high for its entire length. When the two players have succeeded in creating an uninterrupted wall, the official will whistle, at which point the players must step away from the wall.

Scoring: Ten points are awarded, total. The official will determine how many points go to each player, based on percentage of the wall completed by that player, overall quality of the wall, and if any detractions must be made for insufficient height.

FH

farm games!

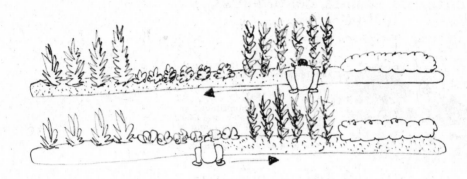

FH

SUPER WEEDING
One on One

The Super Weeding event combines speed and accuracy. Athletes compete to quickly and accurately clear plants classified as "weeds" from 30 m2 of dense, mixed flora.

Procedure: Each athlete is given his/her own area of 5 x 6 m of mixed flora, in which ca. 40% of the flora is classified as "weeds." At the sound of the starting gun, the athlete commences to remove all the weeds from the 30 m2 as quickly as he/she can. When finished, the athlete gets up and returns to his/her starting position and the clock is stopped.

Scoring: This is a timed event, and the athlete with the fastest time (inclusive of penalties) is called the winner. A 5 second penalty is assessed for either: a) each "weed", not removed from the playing area, or b) each non-weed plant removed from the playing area. 2 second penalties are assessed for either: a) incomplete removal of a "weed" (i.e., a portion of the plant is still visible above ground—as per the official's discretion), or b) incomplete removal of a non-weed (as in part a).

Sample Scorecard for Super Weeding:

NAME: Guargle

Competition time: 25:20.8

5-second Penalties: 12
2- second Penalties: 17

Adjusted Final Time: 26:54.8

CHARADES, REBUSES, CONUNDRUMS, ENIGMAS, etc.

(For answers, see page 120)

I

II

I am composed of 25 letters.
My 12, 4, 1, 18, 20, is a word meaning empty.
My 11, 3, 22, 9, 12, 13, 10 is a kind of monk.
My 2, 24, 17, 5, 19 is a piece of money.
My 6, 16, 8, 22 is a division of time.
My 25, 3, 10, 25, 7 are sometimes bad.
My 17, 21, 5 are thought after dinner.
My 13, 3, 14 is large.
My whole is an old saying.

III

What bird is that whose name represents nothing, twice yourself and fifty?

IV

My first in cities is well known,
And by me many live,
Obtain their freedom in the town
And then a vote can give;
My second we can never see,
Whether on the land or sea;
My whole the sailor ofter braves,
When he plows the briny waves.

V

VI

Why does a man in paving the streets correct the public morals?

VII

Entire, I am a companion; beheaded, a verb; replace my head, curtail me, and I am found in nearly every house; curtail again, I am a nickname; reversed, a verb.

VIII

My 1-2-3 designates abbreviations of three states. The whole these states would be before the Revolution. See?

IX

There is a word of five syllables —take away the first and no syllable will remain.

X

I am found in a jail; I belong to a fire;
And am seen in a gutter abounding in mire;
Put my last letter third, and then 'twill be found; I belong to a King, without changing my sound.

XI

My first is irrational, my second is rational, my third is mechanical, and my whole is scientifical.

XII

What word is that to which if you add a syllable, it will make it shorter?

XIII

XIV

What is that which is lengthened by being cut at both ends.

ANSWERS TO CHARADES, ETC.

(1) A man cannot gather grapes from thistles. (2) A penny saved is a penny earned. (3) Owl (O + UU + L). (4) Tradewinds. (5) W HAIR over each eye (I) n xander or a bound will p over t and v iee beef hound. (Where over-reaching and error abound, will poverty be found.) (6) He is amending the public ways. (7) Mate, ate, mat, ma, am. (8) Colony. (9) Horsemanship. (12) Short. (13) Hew hop lace S C on F I dents in awl purse on swill short L y C on F I D E in no body. (He who places confidents in all persons will shortly confide in nobody.) (14) A ditch.

farm games!

REBUSES *by* LAUREN MARESCA

1. To cure your cold, aid your joints, bless your gut, and more and more consume:

2. To prepare your grains for a morning delight, make sure that they are:

3. Oh, what a glorious alchemy is accomplished when one has made:

4. How can you best preserve your cabbage and make it even better for you?

ANSWERS TO REBII ON PG. 173

farm games!

5. Eat enough high-quality: It'll keep you strong!

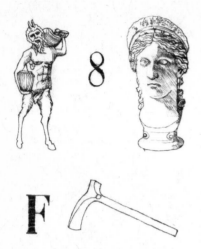

6. "White blood", a.k.a. milk is best consumed:

ANSWERS TO REBII ON PG. 173

A RACE AROUND THE HORN

Two famous clippers set out on a race around Cape Horn in the Autumn of 1852. *Flying Fish* was the winner. On February 1, 1853, she arrived at San Francisco from New York, in 92 days, 4 hours. The loser, *John Gilpin,* had hung on stubbornly, often within hailing distance, but did not end the voyage until 16 hours later.

Although the time was not as good as that set by a famous contemporary, *Flying Cloud,* in 1851, nor by the *Andrew Jackson,* in 1855, the race nevertheless dramatized as never before the 15,000 miles that lay between New York and San Francisco by the ocean route, and the need for faster, surer communication between the two cities. Altogether, only 18 clippers made the run in less than 100 days. And steamers rarely can do better, even now.

Neighbors

ZUCCHINI FLUTE *by* JEREMY SMITH

A Graphic Guide to Making a Zucchini Flute.
(with thanks to Merry from Grassroots Garden in Eugene, OR for teaching so many)

1.   Find a Zucchini plant.

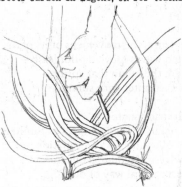

2.   Select a large stem of appropriate shape. Cut the stem off close to the base.

3. Cut the leaf off where it branches out of the stem.

4.   Use the back of your knife to scrape away the spines from the top 2-3 inches of the stem.

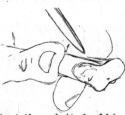

5. Look at the end, it should be roughly heart-shaped. Make sure that the end is solid and that you haven't just created a straw. Find the main vein that ends in the dip in the heart.

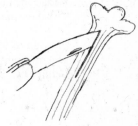

6.   Use the tip of your knife to cut a 1-1 ½" slit in this vein, stopping when you reach the solid end.

7.   Fully cover the slit with your mouth and blow steady but not too strongly. Enjoy the sweet melodies you can produce.

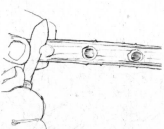

8.   Cut finger holes for making different notes.

9.   Teach others, especially children, this is one of the important things in life.

We find that summer squash work best. Shortening the overall length of the stem creates higher tones, while increasing the length of the slit lowers the tone. We haven't experimented with other members of this plant family; Giant pumpkin bassoons? Cucumber piccolos?

Inadequacies Of Summer

by AVA LEHRER

As a still thing trusts the movement of its parts,
these corrugated waves move motionlessly—
an eternal green, the hymn that springs
the illustrious encounter with summer.

Illustrious, but inadequate
as the need to name it, Inadequacies of Summer,
abrupt erasure of winter's warm isolation,
insistence of skin, and white recollections
tossed out like a natural way of dealing.

This shift that takes as the days
nudge the anatomy of other lesser months.
The day is more than this, than what one remembers—

or is it a different day in a different season
of a different year one thinks of
behind her home, sitting without any knowledge of another.

AGRICULTURAL SONG.

Plough deep to find the gold, my friends,
  Plough deep to find the gold!
Your farms have treasures rich and sure,
  Unmeasured and untold.

Clothe with trees New England's hills,
  Her broad fields sow with grain,
Nor search the Sacramento's rills
  For Californian gain.
Our land o'erflows with corn and bread,
  With treasures all untold,
Would we but give the ploughshare speed
  And DEPTH to find the gold.

Plough deep to find the gold, my friends,
  &c.

Earth is grateful to her sons
  For all their care and toil;
Nothing yields such large returns
  As drained and deepened soil.
Science, lend thy kindly aid,
  Her riches to unfold!
Moved by plough or moved by spade,
  Stir deep to find the gold!

Dig deep to find the gold, my friends,
  Dig deep to find the gold!
Your farms have treasures rich and sure,
  Unmeasured and untold.

———

THE FARMER'S SONG.

"His wants are few, and well supplied
  By his productive fields;
He craves no luxuries beside,
  Save what contentment yields.

More pure enjoyment labor gives,
  Than fame or wealth can bring;
And he is happier who lives
  A farmer, than a king."

———

THE IRRESOLUTE, UNDECIDED, AND THEREFORE UNSUCCESSFUL MAN.

All his defects and mortifications he attributes to the outward circumstances of his life, the exigencies of his profession, the accidents of chance. But, in reality, they lay much deeper than this. They are within himself. He wants the all-controlling, all-subduing will. He wants the fixed purpose that sways and bends all circumstances to its uses, as the wind bends the reeds and rushes beneath it. — *Kavanagh, by Longfellow.*

———

SIMPLICITY.

In character, in manners, in style, in all things, the supreme excellence is simplicity. — *Ibid.*

SUBJECTS FOR DISCUSSION

What kind of fowl is most profitable?

How to make hens lay through the winter.

Should every farmer grow vegetables for home consumption?

Which pays the better—a one-man farm, or a farm on which a hired man is required?

Does a farmer's wife derive enough pleasure out of a flower garden to pay her for extra hours expended?

How much good do we derive from radio programs?

What is the most helpful thing you ever heard over the radio?

Why I love this community.

What are the hardest nuts a farmer has to crack?

What can we do to make home the happiest place on earth?

What is your favorite breed of cattle and poultry?

What can be done to reduce taxes?

What profitable business could be started in our community, on small capital?

How can we raise money for the grange?

What keeps the grange growing?

Am I helping the grange?

Why it would pay farmers to co-operate.

Live Programs for the Lecture Hour

Why I became a farmer.

Why I love the grange.

Does it pay to keep a hive of bees?

Does a berry patch pay?

How best to keep down the insect pests.

Animal diseases and remedies.

Crops and fertilizer.

Is the "flivver" the most serviceable auto for the farmer?

Home curing of hams.

The best method of preserving eggs.

What is the ideal kitchen?

Should a farmer's wife learn how to make her own dresses?

Why it does not pay a farmer's wife to make her own work dresses.

What do we know about lightning protection?

Why we should keep our home and buildings painted.

How can we win back members who have ceased to attend the grange?

Should the grange pay dues for those members who are not able to do so?

Do sheep pay?

How to keep the bottom from dropping out of the egg market.

FARM ENLIVENING SUGGESTIONS FROM "LIVE PROGRAMS FOR THE LECTURE HOUR",
A GUIDE FOR FARMER ORGANIZERS, PROGRAMMERS AND GRANGE HALLS,
WITH COUNTLESS SUGGESTIONS OF RURAL GAMES AND VERSES.

LIVE PROGRAMS FOR THE LECTURE HOUR

Co-operate, That's How We'll Do It

MINNIE CHURCHILL

M. C.

1. How are the farm-ers going to find suc-cess up-
2. How can we get the gov-ern-ment to help us,
3. Where shall we go to ask ad-vice or help at

on the land? How are the farm-ers going to be a
an-y-way? What shall we do to bet-ter our con-
an-y time? Where shall we turn for aid a-long the

great, con-tent-ed band? How shall they meet each prob-lem
di-tions ev-'ry day? How can we help our homes, and
ag-ri-cul-ture line? Where are we sure to find the

that may come in fu-ture years? What be their mot-
keep the boys up-on the farm? There is but one
best as-sist-ance for our cause? Where, but our own

LIVE PROGRAMS FOR THE LECTURE HOUR

Co-operate, That's How We'll Do It

THE HAPPY JOURNEY

SHAKER HYMN
PUB. 1801

ARRANGED BY
ADRIENNE YOUNG-RAMSEY

Oh the happy journey that we are pursuing
Come bretheren and sisters let's all strip to run
Let all be awakened and up be doing
That me attain to our destined home

The heavens of glory is our destination
We're swiftly advancing to that happy shore
We are traveling on in the regeneration
And when we go through we shall sorrow no more

This beautiful journey that we've undertaken
Exceeds all the travels that ever has been
And those that perform it will never be shaken
Because it leads out of the nature of sin

ANSWERS TO REBII FROM PAGE 164

1. Beef, chicken and lamb stock
2. Soaked overnight
3. Sourdough Bread
4. Fermentation
5. Raw and/or cultured
6. Saturated fats

Hoist flags or billowed sails
fireworks sinking, anchored
to find a place to swim.
Fried egg on roof of truck

–Francesca Capone

JULY

———

HAIL

JULY 2013

zodiac signs

Cancer — Most Fruitful — Good for planting, irrigating, grafting, transplanting

Scorpio — Fruitful — Exceptional for vine-type growth

Pisces — Fruitful — Good for root crops & any poor time to make transplants

Taurus — Semi-fruitful — A time to plant brocoli, cabbage, sturdy stalks – root crops

Capricorn — Semi-fruitful — Good time to plant tubers - root crops

Libra — Semi-fruitful — Time for seeding of hay, grain, flowers and crops where beauty is wanted

Aries — Semi-barren — Time for cultivating – plowing where weed-destruction is sought often. Chickens hatched now will be poorer

Sagittarius — Semi-barren — Best for destroying unwanted plant life

Aquarius — Barren — Suitable for seeding – harvesting of unwanted growth

Virgo — Barren — Good for bringing order to the garden – general cleanup

Gemini — Barren — Great to cultivate ing – destroying unwanted growth

Leo — Barren — A time for killing weeds – unwanted growth. Color should be chosen off for change – A good time to cut hay – if earth is too dried, or late in the increase of the moon

---

1 — In 1895 a new Division of Agrostology was established which conducted research on grasses and forage plants.

2

3

4

5 — Carl Einstein dies. *"Where the Column advances, one collectivizes. The land is given to the community, the agricultural proletarians, slaves of caciques which they were, metamorphose themselves as free men. One passes from agrarian feudalism to free Communism."*

6

7

8

9

10

11 — 1966: Lou Gottlieb, founder of the Morning Star Ranch (fondly known as the "The Digger Farm") dies. Gottlieb formerly of the folkie Limelighters, along with Ramón Sender, opened the 32-acre Ranch to anyone who wanted to live there.

12

13

14

15

16

17

18 — 1914 -Seattle holds its first Potlatch festival, including bombing the city with flour bags.

19

20

21

22 — Plant the bean when the Moon is light; Plant potatoes when the Moon is dark.

23

24

25

26

27

28 — In 1900, the Hamburger is created by Louis Lassing in Connecticut.

29

30 — In 1867, Congress sets up Peace Commission with three stated objectives: (1) to end Indian Wars by giving them whatever they wanted; (2) to make peaceful farmers of them; & (3) to get their permission to build railroads across the plains.

31

PROFITS OF FARMING.

Although farming does not lead to the rapid accumulation of wealth, yet it yields a competence to the industrious and frugal, and is the most independent calling in life. The items of profit, though apparently small in themselves, amount, when those for the year are added together, to no inconsiderable sum. And a farmer's expenditures in money being less than is required by other kinds of business—nearly every article for the support of his family being raised on his farm—his balance-sheet shows a healthy and thriving business at the close of the year. Commerce and manufactures are the bases of large cities, and the sources of rapid accumulation of wealth, the legitimate products of which are luxury and excess. These crept into the Roman Republic, and undermined the very foundations on which it was established—industry, morality, and virtue. Licentiousness and vice of every kind followed in their turn, corrupting the better portion of her people, and in due time did their work of destruction. Thus terminated the existence of one of the most magnificent political structures of which the old world could boast.

"It is probable that within the next hundred years our methods of generating a power supply will have to undergo a radical change. Both oil and coal exist in limited quantities, figst of nature which we are consuming without regard to our future needs."

-The North American Almanac, 1927

FIG. 23

Copper plate prints borrowed and printed from Rogersons Hardware Store, Hudson, NY

Woman's Labor in the Field.

In this country it has been a rare sight to see women engaged in field labor, except at the South, where woman's labor has been highly appreciated, especially for certain kinds of work. The present scarcity of laborers leads many to employ women in field labors. During the month of June thousands have found profitable employment in weeding carrots and mangels, setting out cabbages, tobacco, etc., lending a hand in the hay field, and perhaps in the corn and potato fields too. They are paid 50 to 80 cents a day, and we have no doubt they earn it well. At least their employers are entirely satisfied. We should be very sorry to see the women of America subjected to the cruel drudgery of the women of Europe, yet no one can look upon this out-of-door labor, if not of a character to overtax their strength, as likely to work any thing but good to those who participate in it.

Hundreds of **farmers boys, ambitious to do** " the work of a man," and encouraged in it by their parents long before they have man's strength and endurance, have been stunted for life, dwarfed, or drawn out of shape, and still remained healthy and strong, while others have contracted disease, lessening their ability, and shortening their lives. Of course females are quite as likely as males to injure themselves in the same way. American women, and women folks of the farm not less than others, are proverbially " delicate," nervous and weak. Could the ruddy and brown complexions gained in the field, become fashionable, and the " interesting" pale-faces of the darkened parlor find themselves decidedly in the shade, the next generation would have an additional reason to be grateful to this, and to these cruel war times.

Demeter, goddess of the harvest

KWR

EDITORIAL NOTE: The Women's Land Army was formed to provide farm labor during WW1 and WW2. It was an amazing undertaking, and many of the organizers were leading suffragettes, later involved with starting women's colleges and other institutions of civic good. They designed uniforms, recipe books, songs to sing, efficient ways of moving food in dormitories. The ranks within the women's land army were not based on class background-- or so they said. Regardless, it is a beautiful history worth revisiting as we consider the project of mobilizing another land army, of entrepreneurs. - SvTF

The Abandonment Of The Countryside
(After Raymond Depardon's Country Profiles)

by JONATHAN SKINNER

I will never get up at six in the morning to milk a cow
I will never be passionate about farm animals
I will never marry the girl next door
I will never negotiate the sale of a calf over half a bottle of wine
I will never live with the reek of animals in my yard
I will never know not ever leaving my village
I will never hold the light of the Cévennes in the palm of my hand
I will never grow my hair out long and speak to no one
I will never drive my tractor into the village for groceries
I will never experience the sorrow of a cow who won't get up
I will never deliver a sermon on a mountainside
I will never get used to a bare kitchen
I will never live in the house where
I was born I will never talk to my sheep in Occitan
I will never pat a cow on the ribs as I am milking her
I will never fondle a tractor the way I do my car stereo
I will never finish school at sixteen
I will never live in a village of four called Vil Aret
I will never wear a French man's v-neck sweater and cap
I will never grow old a bachelor in love with sheep
I will never sell off my animals feeling I've died
I will never have the satisfaction of haying my twenty cows
I will never place a personal ad in an agricultural rag
I will never discuss the neighbor's inheritance around the kitchen table
I will never have knuckles knobby with farm work and cold hours watching flocks
I will never live to see all of my neighbors die or move away
I will never get used to the rhythm of cow bells
I will never get back to the land

MN

The Declaration of Energy Independence of These New England Citizens

By DORN COX

When in the course of human events it becomes necessary for a people to free themselves of their dependence, they should declare the causes which impel them to the separation. We the undersigned have come to recognize the undue influence of imported petroleum on our lives, and hereby resolve to rectify the situation.

Prudence and common sense dictates that we must act now to begin a transition away from dependence, and towards independence. This goal is desirable to all citizens of our country, and even to the world, to reduce a primary cause for military conflict and suffering. When a society allows itself to become critically dependent upon a limited resource, whose increasing demand and decreasing supply is a guaranteed source of continued conflict, suffering and hardship, it is their right, it is their duty to free themselves, if they have the means to free themselves of that

dependence. Such necessity now falls upon us to undertake efforts toward that goal.

Because oil fueled our economic expansion and enabled us great wealth as a region, over the course of the past millennium the tyranny of petroleum over our way of life and culture has grown, and in the spirit by which these great states were founded we do now find it obligatory to take a stand and restore the independence of our communities from this foreign control. We also recognize and call out in the spirit of free enterprise that the petroleum industry runs contrary to that spirit, being a finite resource influenced by cartels, supported by vast expenditure of tax dollars to guarantee that supply.

These transgressions upon our communities are numerous and egregious.

• We have been required to sacrifice our sons and daughters in the defense of this supply.

• We have unduly spent our tax dollars supporting unrepresentative regimes that plot to harm us, in efforts to maintain these questionable alliances for maintaining the supply.

• We regret the money spent widening our trade deficit to other nations, weakening our economy, and lessening our investment in technology and domestic industry.

• We regret the uncompensated costs to our farmers and agriculture and fishing industry which have been harmed.

• We regret the inequity of market prices not set by free market forces or prices which are unrepresentative of the social and economic cost to our communities at home.

We do herby declare that we will use every innovative and creative means and device to free ourselves from the shackles of this oppression and encourage and employ the industry, experience, and natural and renewable resources at our disposal to overcome and eliminate our dependence. We believe that it is in the strategic interest and defense of our states and communities to control and produce the resources that give our economies motion. Our region is endowed with vast intellectual resources with a history of innovation embodied in our schools and great universities. We call upon ourselves to employ them to the greatest advantage and overcome this obstacle.

Our means will include but are not limited to:

• Invest and commercialize existing technologies, and support research of new technologies

• Shift investments and research away from fossil fuels and towards renewable fuels and efficiency technology.

• Transform our municipal and farm waste streams from costs and liabilities into assets and renewable energy feedstocks applying existing and new technology.

• Work tirelessly to develop locally produced biofuels that fit with the regional climate and into sustainable practices.

• Shift and monetize defense, health and environmental externalities of petroleum dependence onto petroleum users.

• Improve our fuel economy, and increase our home efficiency.

• Upgrade our public transport, and revitalize our downtowns

• Educate and disseminate to our communities these facts and solutions and engage in active discussions across our towns, cities, farms and industries.

With these changes we can harness the energy, spirit, and courage of our heritage and revitalize our towns and countrysides and can reclaim the promise of our land.

We declare that this is hereby the foremost important initiative for our generation . Our resolve and our success or failure will be felt by generations to come. We call to re-invigorate this spirit, and strive towards achieving the goal of complete energy independence. We declare that if we are possessed with the ability to be energy independent then we have a duty to provide the greatest possible security for our economy, our communities, and the generations yet to come.

While we support open trade and free enterprise, we also recognize the importance of security to our citizenry to not be beholden to a foreign supply of the very power that gives our society and economies motion. Let us make the means with which we produce our goods, heat our house or go to work, our daily tribute to the communities and the land we love.

We posses the technology and ingenuity to achieve this goal – we now resolve that we also possess the will to do so. We resolve not to live forever as dependents, but to throw off those chains and become independent. Let us not only object and reject, but let us lead. Let us lead our towns, our cities, our counties, our country, our nation and our world.

GW

Agriculture In A Steady State Economy

by DEBORAH SCHOBER

OVERGROWN AGRICULTURAL SYSTEMS

Since the dawn of the industrial revolution, the scale of farms has increased immensely. Artificial fertilizer, tractors, and other inventions, coupled with cheap oil, have simultaneously increased food production and decreased costs. Large-scale food production has lowered prices to the extent that U.S. citizens spend only 11% of their income on groceries, compared to roughly a third fifty years ago. Measured on a dollars per calorie basis, big agribusiness appears to be efficient. The problem, however, is that low food prices externalize many costs. The costs of degraded ecosystems, resource shortages, and fractured farm communities are not reflected in these low food prices.

FOR EXAMPLE:

• Confined animal feeding operations supply markets with cheap meat, but the price of the meat does not include costs like environmental harm from toxic lagoons or the emergence of drug-resistant microbes from the overuse of antibiotics.
• The price of grain from intensively irrigated cropland does not account for the ecological damage caused by water diversions or the loss of future water supplies from the depletion of aquifers.
• Costs for shipping food, applying chemical fertilizers, and operating farm machinery do not include direct pollution costs or the costs of climate change associated with production of greenhouse gases.
• Low food prices from outsized farms also do not reflect the transfer of wealth from farm communities that practice good land stewardship to corporations that manage lands based on short-term profitability.

Real efficiency in a system (including an agricultural one) is measured by energy return on energy invested. Smaller, less mechanized farms produce more calories of food per calorie of energy expended to grow the crops.

FEATURES OF AGRICULTURAL SYSTEMS IN A STEADY STATE ECONOMY

Agriculture and economics are thoroughly intertwined. Economists like Francois Quesnay and Adam Smith recognized that agricultural surplus "frees the hands" for the division of labor and provides the origin of money. Just as a growth economy tends to impart industrial characteristics to its agricultural systems, a steady state economy tends to impart sustainable characteristics to its agricultural systems.

A steady state economy is characterized by stable or mildly fluctuating population and per capita consumption. Such an economy requires a fixed quantity of food. There is no need for constantly increasing the amount of food produced, and there is a calming effect on the landscape – not as much land needs to be in crop-production mode. In addition to stable population and consumption, a steady state economy features stable and relatively low throughput of energy and materials, a characteristic that applies to the agricultural sector.

The best way to achieve sustainable throughput in agricultural systems is to decentralize. Inputs, especially fossil fuel inputs, can be reduced by shifting to local systems of production, distribution and consumption. Large-scale agribusiness contributes a significant percentage of all fossil fuel emissions in the U.S., stemming from energy-intensive methods of planting, fertilizing, harvesting, packaging, and distributing food supplies. With smaller-scale, more sustainable practices, there is: less reliance on fuel to run heavy farm equipment for production and irrigation; less application of pesticide, herbicide, and fertilizers; less reliance on long-distance transportation to ship crops to processing plants and supermarkets; and less use and disposal of plastic

BENEFITS OF FARMING IN A STEADY STATE ECONOMY

Increased Food Security and Quality

In a study of 200 small-scale farms around the world, agronomist Jules Pretty found that farmers using sustainable practices had an average increase of 93% in output per hectare. In addition, strong relationships between consumers and growers provide greater incentives for quality and safety.

Improved Ecological Health

Farmers and families that double as land stewards understand their immediate environment better than absentee corporate owners. Small-scale farms tend to have greater crop diversification, fewer problems with invasive plant species and harmful insects, increased soil fertility, and more habitat for wild species.

Opportunities for Meaningful Employment

A national agricultural system consisting of local and ecologically sound farming provides numerous job opportunities. Such farming jobs require skillful application of techniques using the mind and the hands.

Vibrant Local Food Economies and Healthier Farm Communities

Community supported agriculture, farmers' markets, and other direct farmer-to-consumer interactions are demonstrating the vitality of local food systems. Wealth is reinvested in the local economy, promoting healthier communities. Relationships are fostered between producers and consumers, creating a greater sense of community and improving quality of life.

New Economics Institute based in the Berkshires , formerly called E.F. Schmacher Society,
they have hundreds of lectures, podcasts, videos on new economic theories.

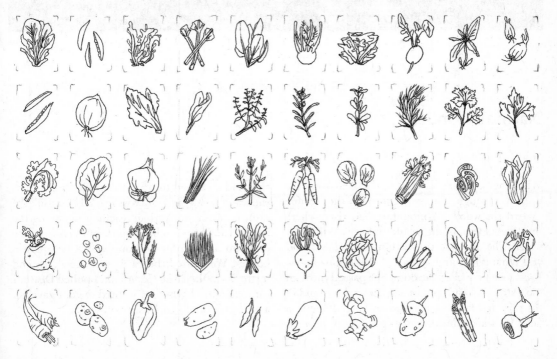

BB

AML

Natural Wonder

by BEN FAMA

When her eyes moved I saw the field too. We walked through the garden, over the cobble toward the woods. A mountain had risen where the hideout was. Branches lifted slowly, as if under water. The universe is breathing she said. I said yes. What is it saying she said. I said I don't know. I was sure I'd locked the door. She passed through too. After the road split we were forced to stand in the rain and later we reached an open circle in a field. We walked inside my little hut. I saw how my skin had absorbed the dye off my clothes. I knew she was a witch. She sat cross-legged and took the tarot out from my vitrine, reaching in among the feathers and bones, a little bottle of water from the fountain of youth in Collioure, near the Spanish border. We shifted the cards in waves back and forth. She pulled The Hangman, I pulled Death. I turned on the TV but she wanted to go. I gave her a music box and I think she still has it.

Dynamite And Agriculture

by JEN GRIFFITH

In 1912, E. I. du Pont de Nemours and Company published a Farmer's Handbook offering new means for breaking up hard-pan and subsoils, clearing stumps, planting orchards, plowing, digging post holes, and regenerating worn out farms. The technological innovation was dynamite. And, the publisher of the handbook is more commonly known as DuPont, the American chemical company and producer of Red Cross dynamite. Farmers could write away for a free copy of the handbook which touted the uses for dynamite in agriculture.

Ordinarily plowing merely turns over the same old soil year after year, and constant decrease in crops is only prevented by rotation or expensive fertilizing. With "Red Cross" Dynamite you can break up the ground all over the field to a depth of two or three feet, for less than the cost of adequate fertilizing, and with better results.

From 1911 to 1913, DuPont sponsored a study dynamite's effectiveness in cultivating fields. Data showed no significant differences in crop yields, soil moisture, nitrate levels, or bacterial activity on the blasted soil. Two years later, researchers found the structure of the dynamited soil was compromised, described as "compacted and puddled" where the dynamite had exploded.

After this study, DuPont advertisements omit dynamite's role in regenerating soil and emphasize its benefit as a labor saver. During WWI, a company ad explains how dynamite has allowed agriculture to continue despite the labor shortage due to men being shipped overseas. According to the ad, in 1917, farmers bought approximately 25,000,000 pounds of dynamite using it for clearing land of stumps and boulders, draining wetlands, breaking up subsoils, excavating cellars and ponds, planting fruit trees, clearing mud from ponds, grading roads, among other labor intensive tasks.

North American farmers were a strong market for the dynamite industry and it became a popular tool. It was so accessible that children set it off as cheap fireworks and could be ordered from Sears by the caseload. Some land grant universities offered courses on use of dynamite in agriculture since it was such a common practice.

Companies also researched and advertised uses for dynamite in animal slaughtering. An article in the September 3rd, 1887 Quebec Daily Telegraph describes an experiment in how to humanely kill livestock with a small stick of dynamite, wire, and an electrical fuse fastened to the animal's forehead. Noble's Explosive Company, the developer and original producer of dynamite, developed this slaughter method.

The chemical companies producing dynamite had robust research departments developing practical uses for the product. Through advertising, these ideas spread throughout the farming community. This trend has repeated itself with other military technologies. Following WWII, the US government found a home for the surplus of ammonium nitrate as a fertilizer for farmers. More recently, Global Positioning Systems used the Gulf War are being marketed to large scale farmers as a resource for decision-making by using satellite imagery and geospatial tools.

Current purchase of dynamite requires a federal explosive license. Chemical companies and land grant universities no longer recommend the use of dynamite in agriculture.

FIG. 86.— Strawberry blossoms.

In shape of breezes, red
earth of oceans like a bathtub
gulf or sound hot wood
patio with bare feet

–Francesca Capone

AUGUST

HURRICANE

AUGUST 2013

zodiac signs

Cancer	Scorpio	Pisces
Sensible plant-ing, irrigating, grafting, transplanting. Most Fruitful	Exceptional for vine-type growth. Fruitful	Good for root crops but a very poor time to make abundance. Fruitful

Taurus	Capricorn	Libra
A time to plant lettuce, cabbage, sturdy stalks, root crops. Semi-fruitful	Good time to plant tubers, root crops. Semi-fruitful	Time for seeding of hay, grain, flowers and crops where beauty is concerned. Semi-fruitful

Aries	Sagittarius	Aquarius
Time for cultivating, plowing where weed destruction is sought after. Chickens hatched now will be potent. Semi-barren	Best for destroying unwanted plant life. Barren	Suitable for seeding, destroying of growth. Barren

Virgo	Gemini	Leo
Good for bringing order to the garden, general cleanup.	Great for cultivation, destroying unwanted growth.	A time for killing weeds, unwanted growth. Color should be shown off for storage. A good time to cut hair if early hair is desired, cut hair in the increase of the moon.

8 ♌ TH ☍☽ ☉ 4:00 20:03

16 ♍ ♐ F ♌ ☉ 6:08 19:52

24 ♓ SA ♌ ☉ 6:15 19:40

1 ♉ TH ☍☽ ☉ 5:53 20:11

9 ♌ ♍ F ☍☽ ☉ 6:01 20:02

17 ♐ SA ♌ ☉ 6:09 19:51

25 ♓ ♈ SU ♌ ☉ 6:16 19:39

2 ♉ F ☍☽ ☉ 5:54 20:10

10 1931 · US: Unemployed Citizens Leagues (UCL) are organized around Seattle during this month. The UCL locals cut wood for fuel from vacant land, harvested unsold crops, planted gardens, & caught fish, all with volunteer labor. ♍ SA ♌ ☉ 6:02 20:00

18 PG ♐ SU ♌ ☉ 6:10 19:49

26 ♈ M ♌ ☉ 6:17 19:37

3 AG ♊ SA ☍☽ ☉ 5:55 20:09

11 ♌ ♍ SU ♌ ☉ 6:03 19:59

19 ♑ M ♌ ☉ 6:11 19:48

27 ♈ ♉ TU ♌ ☉ 6:18 19:36

4 "Attending to the lunar rhythms should refine and enhance one's practices, never paralyze or unduly postpone one's work by waiting for only the most perfect time" -Stella Natura ♊ SU ☍☽ ☉ 5:56 20:08

12 ♍ M ♌ ☉ 6:04 19:58

20 If the full moon rise pale, expect rain. (FULL) ♑ ♒ TU ♌ ☉ 6:11 19:46

28 ♉ W ♌ ☉ 6:19 18:34

5 ♊ ♋ M ☍☽ ☉ 5:57 20:05

13 The Perseid Meteor Shower peaks tonight and tomorrow night, radiating up to 60 meteors every hour. Look to the northeast once the moon has set, after midnight. ♍ ♎ TU ♌ ☉ 6:05 19:56

21 1831 · Nikolai Gogol visits the printers to oversee the production of Evenings on a Farm, finds the typesetters greatly amused by the proofs, & concludes he will be *"an author entirely to the taste of the common people."* ♒ W ♌ ☉ 6:12 19:45

29 ♉ TH ♌ ☉ 6:20 19:33

6 (NEW) ♋ TU ☍☽ ☉ 5:58 20:05

14 ♎ ♏ W ♌ ☉ 6:06 19:55

22 ♒ ♓ TH ♌ ☉ 6:13 19:43

30 ♉ ♊ F ♌ ☉ 6:21 19:31

7 ☍☽ ♌ W ☉ 5:59 20:04

15 ♏ TH ♌ ☉ 6:07 19:54

23 ♓ F ♌ ☉ 6:14 19:42

31 ♊ SA ♌ ☉ 6:22 19:29

ON ELI WHITNEY

In 1793 it was completed and thereupon misfortune...Scanty was the gratitude of his generation, however, that of posterity has been profound. The cotton gin paid off the debts of the South, increased its capital and tripled the value of land. It was a source of fortune to thousands a boon to all mankind, a perpetual and priceless memorial to the name of Whitney.

-The North American Almanac 1927

THE NEW OLD TRADITIONS THROUGHOUT THE YEAR

SWELTER: "Change Can Make You Sweat" (On/around August 1st)

Sometimes a person can have too much of a good thing. Although the Sun shows humility in passing its influence over to the Moon, change often comes with difficulty and it is by way of this difficulty that we experience the sweltering heat of Summer's end. So powerful is the Sun in its influence that even as the Moon begins to walk toward dominance, the heat grows hotter. Swelter is a time to celebrate what does and does not serve us.

It Is August That Brings

by DOUGLASS DECANDIA

it is August that brings the communion of Summer and Autumn.

these are the days of sweat and sweaters, of cool morning and warm noon.

the fruit has been set and the Will to ripen strong upon the plant and in the belly of the collector.

the field grows in deep color as the forest becomes a pallet for the earth's brush.

seeds fall from broken pods and roots grow stronger as Summer calms and the Wild Ones fill upon her bounty.

these are the days when our body begins to slow and the cool airs clean the chambers of our mind.

ATH

EDITORIAL NOTE: These presses are used to press sugary sap from the stems of sorghum plants. This syrup is then boiled down into either a syrup or molasses consistency. This year, across the midwest, terrible drought was suffered by thousands of farmers across millions of acres of feed corn, mostly GMO, and it deeply impacted yields. Corn prices soared as yields were pitiful. Meanwhile, sorghum crops, which bear quicker, tolerate less rainfall, and can be planted later in cases of wet spring conditions, thrived. Mark Twain suggested, "If you put all your eggs in one basket, watch that basket." Our nation's eggs are very much bound up in a small handful of crops, all grown in monoculture, and many of them genetically modified to resist herbicide, not climate change. We'd do well to heed his advice and watch for opportunities to plant more resilient crops, to diversify, to build soil carbon and moisture holding capacity with winter cover crops, even (imagine) to start rotating animals back through our crop fields to build soil life. Meanwhile, sorghum is a crop grown for animal fodder, bird seed, cane syrup, and health food markets. It is a good option to access for individual farmers, and the kind of crop-modification, diversification, and rotation planning that takes into account the brittle ecology and ruthless economy of industrial farming. -SvTF

CANE MILLS.

SINCE IT HAS BECOME SO COMMON WITH FARMERS to manufacture their own molasses in the form of sorghum, some good contrivance for expressing the juice has become very desirable. Fig 1 represents a sugar mill of the kind, to be placed over a barrel and driven by hand. The fly wheel makes the work comparatively easy. The hopper is represented as lifted off to show the rollers and gearing. The whole, without the barrel, holds about ninety pounds. The rollers are about eight inches in length, and four in diameter. This mill is mostly intended for common sugar cane.

For expressing the juice of the sorghum by horse power, which is generally adopted, the horizontal mill represented by fig. 2, answers a good purpose, and is one of the best in market. It possesses the advantage of convenience in feeding, with strength, lightness and cheapness. The rollers are about ten inches in diameter. It will be found useful in the South, and all sugar raising countries where only a limtied quantity of cane is ground. The mill is made of three sizes, with wood or iron frames. The cut shows the medium size. The entire weight of this size is six hundred pounds.

Fig. 1.

Fig. 2.

WET FEET.

I have only had three pair of boots for the last six years, (no shoes) and I think I shall not require any more for the next six years to come. The reason is that I treat them in the following manner; I put a pound of tallow and half a pound of rosin in a pot on the fire; when melted and mixed, I warm the boots and apply the hot stuff with a painter's brush until neither the sole nor the upper leather will suck in any more. If it is desired that the boots should immediately take a polish, dissolve an ounce of wax in spirits of turpentine, to which add a teaspoonful of lamp black. A day after the boots have been treated with the tallow and rosin, rub over them this wax in turpentine, but not before the fire. Thus the exterior will have a coat of wax alone, and shine like a mirror. Tallow or any other grease becomes rancid, and rots the stitching as well as leather; but the rosin gives it an antiseptic quality which preserves the whole. Boots and shoes should be so large as to admit of wearing cork soles. Cork is so bad a conductor of heat that with it in the boots the feet are always warm on the coldest stone floor.

boot hack

boots
scuffle
hoe

hilary's scuffle toe

ahhh.

i o n a
MAY 0 1 2012

Pastoral
FROM PRAYERS & RUN-ON SENTENCES

by STUART KESTENBAUM

The fields make a harmony of the rusted
remains of cars, the boulders left behind
by the glaciers, the styrofoam cups
hung up in the underbrush,
the crows overhead. The barns join them
in this song. Empty of use, holding
only memories of cows and hay, they begin
to sag toward earth. On route 3, right after
the new acres of Bank of America
where telemarketers sow debt in the old fields,
the barn stands, like a rock in the river
of wind that erodes it everyday. To think
that wind and water, that famous partnership,
can take this down in a slow dismantling
of purpose, purposeful in itself because it
gets us back to the circle of beginning and ending.
Everything is headed that way, one day,
when it looks like death has just visited,
and there is more collapsed than standing,
the posts and beams that were once the trees
of the forest, fallen a second time,
and we all know what sound that makes,
the sound that rings in the ears of philosophers,
those guys who are always around at the end
trying to make sense of things.
They're out there listening now
to the steady March wind,
which has the cold of February
and the promise of April,
each its own story
told again and again.

Old-time haying

Vermont Sail Freight Project

A carbon-neutral transport initiative connecting the farms and forests of Lake Champlain with the Lower Hudson Valley

GOALS:

To involve students and the local community in enhancing farmers' livelihoods and regional resiliency.

To develop and prove a modern method of cargo distribution using sail power.

To generate excitement and inspiration in Vermont, New York, and nationally about creative twenty-first century energy innovation

VISION:

The Vermont Sail Freight Project is a contemporary re-invention of an historic regional foodway.

In 2013, the VSFP team will build a simple low-cost sailing barge with volunteer and student support. A two-person crew will sail with 8 tons of non-perishable food cargo from Lake Champlain down the Hudson, selling at ports south of Fort Edward. Interactive online maps and a purchasing platform will allow buyers and the public to track the boat's progress to New York City, where it will sell its Vermont-brand cargo at NYC retail prices. There the boat will acquire a new cargo of non-perishable imported food items (such as coffee, sugar, and chocolate) and return to Vermont to distribute them, completing the voyage with at least 10 times more energy efficiency than a semi truck—and considerably more inspirational, conversation-starting mass appeal. A core aim of this initiative is to incubate an economically sustainable transport and distribution model that will become farmer owned and operated within 2-3 years.

THE FIRST VOYAGE: 2013

VERMONTSAILFREIGHTPROJECT.WORDPRESS.COM

Strawberry Fields

-Tensie Hernandez
Guadalupe Catholic Worker
· California

In our area, farm fields stretch on for miles. As you pass them on the roadways they look like an ocean of green. And like the ocean, mysteries lie beneath their surface in the soil. This is agribusiness country located on the central coast of California. Our Catholic Worker house is nestled among these fields in Guadalupe, a town named after the virgin who cares for all its inhabitants.

Our Catholic Worker house was started almost 15 years ago with a deep calling to serve the needs of the Guadalupe community that almost exclusively comprises immigrant field workers. One of the most pressing needs is advocacy for those who have serious illnesses and face a nearly impossible challenge in navigating a medical system filled with bureaucracy and cracks.

As agribusiness goes, this area has all of the trappings that allow inherent injustices to go unchecked. The principle crops have always been broccoli, celery, cabbage and lettuce. These crops thrive with the cool morning fog that blankets the valley most of the year. The workers are immigrants from Mexico and Central America; most are undocumented. With the advent of hybridizing crops, strawberries have also become quite a lucrative business, generating almost $400 million in this area.

Strawberry picking is arguably the most difficult of fieldwork. Oaxacans (from Southern Mexico) are the favored workers for this crop; they are short in stature and more able to stoop 10-12 hours a day in what most consider to be literally back-breaking work.

Fumigants are commonly used by agribusiness in the fields of Guadalupe. The effect of these chemicals is constantly being treated in our free medical clinic. Related complaints range from allergies and headaches, to more serious thyroid conditions. Some pesticides are more deadly than others. Methyl Iodide is among the deadlier ones.

Used in strawberry fields, Methyl Iodide is injected into the soil to literally sterilize it before planting. It was the fallback pesticide used to replace another very toxic and ozone layer-destroying pesticide: Methyl Bromide. Methyl Bromide was legally phased out as a pesticide in 2005. In 2006, Japanese chemical giant Arysta presented Methyl Iodide to the EPA as the perfect replacement for Methyl Bromide. The EPA approved the use of Methyl Iodide the very next year, despite the fact that more than 50 scientists and doctors--including five Nobel laureates in chemistry--warned the agency about the dangers of Methyl Iodide. Among the dangers cited by the scientists is that exposure to Methyl Iodide --which readily contaminates groundwater--causes late term miscarriages. Moreover, Methyl Iodide is so reliably carcinogenic that it's used to create cancer cells in laboratories.

These cancer cells have also been found in the bodies of over a dozen Oaxacan fieldworkers I've accompanied during the last 5 years—each of them suffering from advanced cancer, and all of them strawberry fieldworkers. The vast majority were under 40 years of age. Six have died, leaving small children behind.

The United Farm Workers have begun a campaign to eradicate Methyl Iodide's use. As of this writing two counties north of Guadalupe have passed a resolution urging the governor to re-examine the state's approval of its use. We await the outcome.

Meanwhile, the price for cheap year round strawberries is paid for in the lives of those who harvest them.

Summary Toxicity Information: methyl iodide

PAN Bad Actor Chemical [1]	Acute Toxicity [2]	Carcinogen	Ground Water Contaminant	Developmental or Reproductive Toxin	Endocrine Disruptor
☠	☠	☠	Potential	?	?

 Indicates high toxicity in the given toxicological category.

 Indicates no available weight-of-the-evidence summary assessment.

PESTICIDE ACTION NETWORK IN THE US
LEADING ORGANIZATION WORKING TO PROTECT FARMWORKERS,
WATERSHEDS AND AIR QUALITY FROM AGRI-CHEMICAL POLLUTION
WWW.PANNA.ORG

Hoe

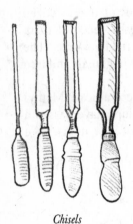

Chisels

Howland Tools

CO-WRITTEN *by* ELISE MCMAHON & SHELBY HOWLAND

I first met Shelby Howland after a scything workshop he was holding at the farming community Katywil, in Western, MA. He is both idealistic and cynical, a modern purist stubbornly supportive of people gaining the knowledge to use hand tools for farming. After starting his own farm using only hand tools at the age of 16 and having such a frustrating time locating a good quality tool made and distributed in America, he began his research in their proper use and manufacturing. He started the business Howland Tools 2 years ago in response to this research, sourcing tools in Europe where hand tools have remained economically important for far longer than in the United States, partly due to a longer persistence of smaller farms. This means that these tools have been designed and manufactured for people whose livelihoods depend on their use. Many of these companies have been in production for hundreds of years. When I asked him what he looks for in a tool he gave me a list of qualities - rugged, simple, easily made by hand, durable, effective, and made out of good steel. When asked what tool above others is most effective and versatile, he says the Peasant Hoe, also known as the grape hoe, which has remained pretty much the same design for centuries. With this tool one can mow some grass, weed, till, furrow, cover furrows, move material, dig trenches, cut down a tree, and even harvest cabbage. One person, one tool, making a farm, nice.

EM

"Find the shortest, simplest way between the earth, the hands, and the mouth." - Lanzo Delvasto

24. Ancient and modern harvesting implements, drawn to scale.

I Made (It) Myself : A Call To Hands

by SCHIRIN RACHEL OEDING

It was about five years ago that I received my first axe. It was a beautiful old Snow & Nealley, with, I was told, decades of history to its name, having been forged in Maine sometime in the 1950s. It was given to me bare, just the bit of the axe without the essential handle, and I weighed it in my hands, cold metal, fish-scaled with the marks of the hammer that had peened it a long time ago. The forester who gave me the axe was a collector of old tools, a mischievous giver of gifts that demanded elbow grease on the part of the receiver, someone who lived by the maxim, "Always be finishing something." The axe, of course, needed a handle in order to be of any use to me. I was planning on felling a few trees and splitting some firewood, and so, it being fall already, I needed that handle badly. "You can get one at the hardware store," an axe-enthusias-

Western Winter Wren

tic friend of mine told me, "but it won't last the season." The grain would be off kilter, running the wrong way, making the handle prone to splitting and splintering. Fine, I thought, I'll make one myself. I wanted something that would do this axe justice. I left the forester's house with the axe bit and a large piece of ash —enough wood for two handles, in case I broke one in the process.

Shaping and fitting the axe handle was a slow business. It was an antediluvian process, and I developed a Luddite's love affair with it. Had I been given the axe complete —sharpened, with a handle— I wouldn't have had the chance to go down this road at all. Always be finishing something, I told myself as I shaved long curls of wood from the handle, always be finishing something.

Whenever I had some spare time, I was in the workshop, shaving and filing, watching my hands as they learned to measure and evaluate their work. My hands and fingers, which had lately been engaged almost exclusively in the work of writing, mostly on a computer, were transformed. Along with blisters and new callouses, there was a reawakened and remarkable dexterity, a muscle memory that could be depended on to remember where the axe handle still needed to be smoothed out or adjusted in some way. Even when my mind wandered from the work, my hands never lost their direction.

It isn't a big deal, really. Humans have been making things by hand since time immemorial —it's a big part of what makes us human, after all. Or is it? Our hands tell their own stories, and lately, I'd venture to say, many of us have reduced ours to unskilled labor. Luddite references aside, when we make something, a tool, a meal, a piece of clothing, we engage with that thing in a way that we never would if it came to us readymade and complete. We might marvel at something that is done, but we wouldn't go on a journey with it. The axe handle I made was a relatively insignificant thing, a tiny chunk of an enormous, expansive forest, but it took on a mythic quality anyway. It told me the story of a tree, from seed, to sapling, to fully grown; felled, bucked, split, a life transformed by my hands —a story that would easily have been erased by the convenience of just buying a handle and getting on with it already.

With the help of a friend, I hung the axe and cleaned up the edges of the blade. The finished axe handle, long and slightly awkward in shape, wasn't too bad for a first attempt. It wasn't perfect, and that was a blessing —I still had plenty of unfinished work to keep me busy. Completeness, it became clear to me, wasn't happiness per se; it was the getting there that really mattered. Because while I shaped the axe, it shaped me.

I've become weary of the word "convenient." There are many so-called conveniences that I wouldn't want to live without, but I know I wouldn't be happy without the inconveniences (Ahem!) of making things myself. Today, having proudly accumulated a small array of beautiful tools, some of them handmade or rebuilt by people I know, I especially enjoy making things for others. I now grow food for a living, so my relationship to the process of hand making, of coaxing a new form from a seed, a piece of wood, a ball of yarn, has evolved, too. It is a give and take: while I might have some power over the outcome of my work, the materials themselves dictate the process. There is an intimate sense of empathy that comes from working with raw materials, because the search for purpose outside of the self is embedded in hand making. When we make something, we take a trip into our materials, we dig deeply into the soil, become familiar with the minutiae of the thing we aim to transform, and imagine it serving its purpose; we get to take a leap outside of ourselves and exercise our imaginations, both in the tactile and in the intangible realms.

The choice is yours: pick up an axe, a needle and thread, a carving knife, collect wild edibles, prepare a feast, cast on some socks, paint a mural onto your walls, hook a rug, plant a garden, forge a blade. Above all, honor your hardworking hands, celebrate failures and successes, and do what you love.

And remember: always be finishing something.

AF

FIFTY FAMOUS FARMERS

JOHN DEERE
1804–1886

JOHN DEERE, one of the pioneer settlers of Moline, Illinois, ex-king plow manufacturer of the world, and third son of William Rinold and Sarah (Yates) Deere, was born at Rutland, Vermont. His father was a native of England. His mother was born in Connecticut, the daughter of a captain in the British army who came to this country during the Revolutionary War.

In 1805 William Deere removed to Middlebury, Vermont, where John attended the common schools and acquired a good ordinary education. Without the knowledge of his mother, he worked for a tanner at grinding bark, and earned a pair of shoes and a suit of clothes before he was sixteen years of age. When seventeen he apprenticed himself to Captain Benjamin Lawrence of Middlebury to learn the blacksmith's trade, which he fully mastered in four years, receiving in the meantime for his services, each year respectively, the sum of thirty, thirty-five, and forty-five dollars.

After a year or two at this work he removed to Burlington, Vermont, where he hammered out by hand the iron work for a saw and oil mill erected at the neighboring town of Colchester, and acquired thereby a local reputation as a mechanic and iron-worker.

He was married, with a young family, and profits were small, but by perseverance and economy the year 1837 found him ready to try his fortune in the great West.

FIFTY FAMOUS FARMERS

In 1838 John Deere built a dwelling house, eighteen by twenty-four feet, and brought his wife and five children from the East. It was then a weary journey of six weeks by stage coach and lumber wagon.

Settled in his little home, though often shaking with the ague, he pushed forward his work. In 1839 ten plows were built, and the entire iron works of a new saw and flouring mill were made, with no help except that of an inexperienced man as blower and striker. The next year a second anvil was placed in the shop, a workman employed, and forty plows made.

John Deere's fame as a plow maker was now rapidly extending, and in 1841 he made seventy-five steel mold-board plows and built a brick shop thirty by forty-five feet. The year following one hundred new plows were added. The tide which was then set clearly in his favor, afterwards bore him steadily on to fortune. In 1843 he took Major Andrus into partnership, enlarged his buildings by erecting a two-story brick shop, added horsepower for the grindstone, established a small foundry, and turned out four hundred improved plows.

In 1846 the annual product had increased to one thousand, and, as time advanced, improvements were made, but the difficulty of obtaining steel of proper dimensions and quality was found to be a great obstacle to the complete success of the business. Mr. Deere accordingly wrote to Nailor & Company, importers, of New York, explaining the demand of the growing agricultural states of the West for a good cast-steel plow and stating the size, thickness, and quality of the steel plates he wanted. The reply was that no such steel could be had in America, but that they would

JOHN DEERE

send to England and have rollers made for the purpose of producing the special sizes of steel. An order was sent, and the steel made and shipped to Illinois.

During the same year, with the view of developing a market nearer home where he could obtain material for his plows, Mr. Deere opened negotiations in Pittsburgh for the manufacture of plow steel, as is shown by the following extract from James Swank's book, *Iron in All Ages:*

"The first slab of cast plow steel ever rolled in the United States was rolled by William Woods at the steel works of Jones & Quiggs in 1846 and shipped to John Deere, Moline, Illinois, under whose direction it was made."

It was in the shaping of the moldboard that Mr. Deere's ingenuity more particularly manifested itself. He was undoubtedly the first man to conceive and put in operation the idea that the successful self-scouring of a steel moldboard depended preëminently upon its shape. The idea was his, and he worked upon it until its correctness was fully demonstrated.

John Deere's practical foresight enabled him to see that his location was not advantageous for a growing business. Coal, iron, and steel must be hauled from La Salle, a distance of forty miles, and his plows must be taken a long distance to market in the same slow and expensive manner. He therefore sold his interest in the business at Grand Detour to his partner, Mr. Andrus, and removed to Moline, Illinois, where there was good water power, coal in abundance, and cheap river transportation.

A partnership was formed between John Deere, R. N. Tate, and John M. Gould, shops built, and work commenced, resulting the first year in the production of seven hundred plows.

WILLIAM DEMPSTER HOARD

It was in these years of the early seventies, apparently, that he had his first vision of what dairying in the West might become. In 1870 the Wisconsin cheese product amounted to less than one million pounds, and this seemed an enormous amount to the people of those days. Then all cheese was sold on the basis of the Liverpool market, and the bulk of it, except the Wisconsin product, was shipped there. New York State and the Western Reserve in Ohio were the greatest cheese producing sections. Mr. Hoard thought if only the prohibitive freight rates to the Atlantic seaboard could be reduced, dairying would increase by leaps and bounds.

It was costing then two and one-half cents a pound to ship cheese in common cars to New York. Mr. Hoard made a trip to Chicago and interviewed the agents of all the different freight lines to the Atlantic coast, but when he told them what he wanted, they laughed at him. The last man he went to see was Mr. W. W. Chandler, agent of the Star Union Line, the first refrigerator line in the United States. When he went into the office, Mr. Chandler wheeled in his chair and said, "What do you want, sir?"

"I represent a million pounds of Wisconsin cheese seeking an outlet on the Atlantic seaboard at rates that will allow us to compete with other cheese-producing sections," replied Mr. Hoard.

FIFTY FAMOUS FARMERS

Board next week, and come yourself to explain its advantages and workings."

"Is there anything else you want?" asked Mr. Chandler.

"Not now," was the reply.

The very audacity of the request seemed to gain its point, because Mr. Chandler said he would do it. This rate was in effect for many years and is practically the rate to-day. Cheese factories then began rapidly to increase in numbers and the cheese business continually improved, until J. Q. Emery, Dairy and Food Commissioner of Wisconsin, now estimates that Wisconsin cheese products for the year 1913 exceeded 190,000,000 pounds, and in the same year the estimate shows 133,000,000 pounds of butter, which together with condensed milk, market milk, cream, and by-products would exceed in value $100,000,000.

From the very first publication of the *Jefferson County Union*, which is to-day one of the best county papers in Wisconsin, he published one or two columns about the dairy, and he constantly urged the farmers of Jefferson County to go into the dairy business.

In 1885 his son, Arthur Hoard, suggested that he believed there was a field for a special dairy paper. His friend, Edward Coe of Whitewater, suggested the name *Hoard's Dairyman*, and the venture was undertaken. From the small, four-page paper of that day *Hoard's Dairyman* has now become the leading dairy paper of the world, with thirty to forty pages weekly and with a circulation of seventy-five thousand. It goes into every state in the Union and into practically all foreign countries. Sir Henry Lennard, who owns a famous dairy farm near London, sent word to Mr. Hoard that all he knew about the dairy busi-

Nature Study

by LIBERTY HYDE BAILEY

"Leaflet I: What Is Nature-Study?", 1904,
The Country Life Movement, This Holy Earth

"NATURE-STUDY, as a process, is seeing the things that one looks at, and the drawing of proper conclusions from what one sees. Its purpose is to educate the child in terms of his environment, to the end that his life may be fuller and richer. Nature-study is not the study of a science, as of botany, entomology, geology, and the like. That is, it takes the things at hand and endeavors to understand them, without reference primarily to the systematic order or relationships of objects. It is informal, as are the objects which one sees. It is entirely divorced from mere definitions, or from formal explanations in books. It is therefore supremely natural. It trains the eye and the mind to see and to comprehend the common things of life; and the result is not directly the acquiring of science but the establishing of a living sympathy with everything that is."

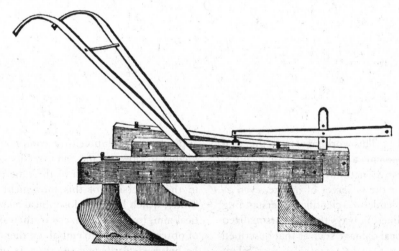

WOOD'S PATENT PLOW CULTIVATOR.—(Fig. 44.)

*The Illustrated Annual Register of Rural Affairs
and Cultivator Almanac for the Year 1869,
Thomas, J.J. (John Jacob) www.archive.org*

Changes in farm labor during the century from 1840-1940 were as great as changes in the techniques of production and the process of marketing. By 1840, indentured labor had practically disappeared from farms, and there was no slave labor. Only free labor remained. Much less man-labor now is required to operate farms, because of the use of labor-saving machinery. A large proportion of it may be considered skilled labor. The operations of modern farming, especially the use of machinery, require a much more alert grade of labor than that which tilled the soil mainly with its brawn. With the development of highly commercialized agriculture and corresponding changes in the rural social structure, less of the work of the farm is done by the farmer's wife, sons and daughters, and more by machinery.

AN EXCERPT FROM PENNSYLVANIA AGRICULTURE AND RURAL LIFE
BY STEVEN WHITCOMB FETCHER,
PENNSYLVANIA HISTORICAL AND MUSEUM COMISSION, 1955

EDITORIAL NOTE: This work grew from the "Nature Study movement" a progressive 19th Century clan of progressive naturalists, whose purpose was to empower the observer of nature, even as the abstract principles of scientific understanding came to be codified in ways that de-personalized nature. Ecological damage such as that described by Rachel Carson, in her seminal work "Silent Spring", was perpetuated by technicians blindly applying one solution to a complex ecosystem, unable to observe and interpret its consequences. Remarkable women of the time participated in the leadership of this movement of Nature Study, which had a philosophical as well as practical aim, to remind citizens of their own powers of observation, and to embolden them to become be better stewards and interpreters of the world around them. -SvTF

LIBERTY HYDE BAILEY
1858–

LIBERTY HYDE BAILEY, who succeeded I. P. Roberts as director of the College of Agriculture of Cornell University, was born on a fruit farm at South Haven, Michigan, March 15, 1858. Mr. Bailey brought to his college work a love and understanding of nature, a power of expression, a capacity for work, genius for inspiring leadership, a radiant and indomitable spirit, and a prophetic foresight, which qualified him in rare degree to take the leadership of a college of agriculture in a great university just at the time when agricultural education needed far-sighted leadership.

The College of Agriculture had struggled for funds and facilities and had fared but meagerly. In most of the states in the Union, the state governments had come to the aid of the agricultural colleges with some funds, mostly small, to supplement the incomes from the original land-grants for these institutions. As yet, New York State had assumed no responsibility for the College of Agriculture in that State, although it had made a few small grants for special purposes. At the outset, Director Bailey persuaded the Trustees of Cornell University to ask the State to recognize the College of Agriculture as a state institution, and to provide for its equipment and maintenance. There was opposition in the State, mostly from other institutions. The fight was a test of strength and of sound state policy for education, and Director Bailey revealed his full powers of leadership,

FIFTY FAMOUS FARMERS

of clear and logical argument, and of compelling conviction. Victory came the first year, when the State appropriated $250,000 for buildings for a New York State College of Agriculture at Cornell University. This was, at the time, a large grant for agricultural education. Two years later, in 1906, after another legislative struggle, Director Bailey accomplished the passage of an Act fully recognizing the College of Agriculture at Cornell University as the New York State College of Agriculture, defining broadly its powers of resident instruction, research, and extension teaching, and placing the administration in the hands of the Trustees of Cornell University as agents for the State.

Then followed rapidly the growth of the college in buildings, land, faculty, and student body. The legislature was generous. Provision was made for the erection of substantial buildings for agronomy, animal husbandry, poultry husbandry, home economics, forestry, a great auditorium, barns and greenhouses, costing with their equipment approximately $800,000 in addition to the first appropriation in 1904. The curriculum was broadened, departments of instruction covering practically the entire field of Agriculture were established, and a staff of persons engaged to develop the work.

He has the art of speaking and writing in phrases that so command the attention of boys and girls, young men and young women that facts and truths of science bud, blossom, and fruit in personal devotion and community action.

Dr. Bailey enjoyed farm life because he had been reared on a farm. At the age of twenty-four he was graduated from the Agricultural College at Lansing, Michigan, and

FIFTY FAMOUS FARMERS

was associated at once with Asa Gray at Harvard, the most scholarly botanist of America.

He was professor of horticulture and landscape gardening in the Michigan Agricultural College for three years, and from 1888 to 1913 he was in horticultural leadership of Cornell University, Ithaca, New York.

Dr. Bailey served as chairman of the world famous "Commission on Country Life" appointed by President Theodore Roosevelt in 1908. This was the greatest opportunity and responsibility for country life improvement that any American has ever enjoyed, and he made the most of both the opportunity and responsibility.

He has edited three Cyclopedias of Horticulture and Agriculture totaling fourteen large volumes. He has written five series of manuals and texts for Rural Life and Rural Schools, totaling more than twenty volumes. He has also written thirty elaborate books on various phases of horticulture, agriculture and country life. No other American has been editor and author of as many volumes on the art and science of activities in field, forest, and farm as has Liberty Hyde Bailey, and no one has written upon any one of his subjects more acceptably than he.

When Dr. Bailey retired in 1913, under protest from faculty, alumni, and trustees, he left a college set far forward in its development, well housed, well manned, and with an impetus and purpose which has never left it. World leaders in Agriculture generally agree that no other nation has had an individual in its country who has done more to promote scientific Agriculture than Liberty Hyde Bailey of New York who has done so much for the United States.

THE BEAVER.

Of all quadrupeds the Beaver possesses the greatest degree of natural or instinctive sagacity in constructing its habitation; preparing, in concert with others of its own species, a kind of arched caverns or domes, supported by a foundation of strong pillars, and lined or plastered internally with a surprising degree of neatness and accuracy.

The *American*, or, as they are called, the *associated* and *civilized beavers*, unite in society in the months of June and July, arriving in numbers from all parts, and soon forming a troop of two or three hundred. If the waters near which they fix their establishment are flat, and do not rise above their ordinary level, as in lakes, they dispense with a bank or dam ; but in rivers or brooks, where the waters rise and fall, they construct a bank, and by this artifice form a pond, or piece of water, which remains always at the same height. The bank traverses the river, from one side to the other, like a sluice, and is often from eighty to a hundred feet long, by ten or twelve broad at the base. This pile, for animals of so small a size, appears to be enormous, and supposes an incredible labour : but the solidity with which the work is constructed is still more astonishing than its magnitude. The part of the river where they erect this bank is generally shallow. If they find on the margin a large tree which can be made to fall into the water, they begin with cutting it down, to form the principal part of their work. This tree is often thicker than the body of a man ; but by gnawing at its foot with their four cutting teeth, they accomplish their purpose in a very short time, always contriving that the tree should fall across the river. They next cut the branches from the trunk, to make it lie level. These operations are performed by the whole community: while some are employed in gnawing the foot of the tree, others traverse the banks of the river, and cut down smaller trees, which they dress and cut to a certain length, to make stakes of them, and first drag them by land to the margin of the river, and then by water to the place where the building is carrying on. These piles they sink down, and interweave the branches with the larger stakes. While some are labouring in this manner, others bring earth, which they plash with their fore-feet, and transport in such quantities, that they fill with it all the intervals between the piles. These piles consist of several rows of stakes, of equal height, all placed opposite to each other, and extend from one bank of the river to the other. The stakes facing the under part of the river are placed perpendicularly : but the rest of the work slopes upwards to sustain the pressure of the fluid, so that the bank, which is ten or twelve feet wide at the base, is reduced to two or three at the top.

THE BEAVER, FROM HANDBOOK OF NATURE
STUDY, BY ANNA BOTSFORD COMSTOCK, 1911

The first great structure is made with a view to render their small habitations more commodious. These cabins, or houses, are built on piles near the margin of the pond, and have two openings, the one for going on the land, and the other to enable the beavers to throw themselves into the water. The form of these edifices is either oval or round, and their dimensions vary from four or five to eight or ten feet diameter. Some of them consist of three or four stories, and their walls are about two feet thick, raised perpendicularly on planks, or plain stakes, which serve both for foundations and floors. They are built with amazing solidity, neatly plastered both without and within, impenetrable to rain, and capable of resisting the most impetuous winds. The partitions are covered with a kind of stucco, as nicely plastered as if it had been executed by the hand of man. In the application of this mortar their tails serve for trowels, and their feet for plastering. They employ different materials, as wood, stone, and a kind of sandy earth, which is not subject to dissolution in water.

These most interesting animals labour in a sitting posture ; and, besides the convenience of this situation, enjoy the pleasure of gnawing perpetually the wood and bark of trees, substances most agreeable to their taste ; for they prefer fresh bark and tender wood to the greater part of their ordinary aliment. Of these provisions they lay up ample stores to support them during the winter ; but they are not fond of dry wood, and make occasional excursions during the winter season for fresh provisions in the forests. They establish their magazines in the water, or near their habitations ; and each cabin has its own, proportioned to the number of its inhabitants, who have all a common right to the store, and never pillage their neighbours. Some villages are composed of twenty or twenty-five cabins ; but such establishments are rare, and the common republic seldom exceeds ten or twelve families. The smallest families contain two, four, and six beavers ; and the largest, eighteen, twenty, and, it is alleged, sometimes thirty. They are almost always equally paired, there being the same number of females as of males. When danger approaches, they warn each other by striking the tail on the surface of the water, the noise of which is heard at a great distance, and resounds through all the vaults of their habitations. Each takes his part : some plunge into the lake, others conceal themselves within their walls, which can only be penetrated by the fire of heaven, or the steel of man, and which no animal will attempt either to open or overturn. They often swim a long way under the ice ; and it is then that they are most easily taken, by at once attacking the cabin, and watching at a hole made at some distance, whither they are obliged to repair for the purpose of respiration.

Beside the associated beavers, there are others which lead a solitary life, and, instead of constructing caverns, or vaulted and plastered receptacles, content themselves with forming holes on the banks of rivers. When taken young, the beaver may be readily tamed ; and in that state appears to be an animal of a gentle disposition, but does not exhibit any symptoms of superior sagacity.

between lathe mornings and maidens
bathing, tomatoes amongst fists
our bodies arch and earn
building relational forms of sustenance

–Leonora Zoninsein

SEPTEMBER

———

FROST

SEPTEMBER 2013

1 ♊ ♋ SU ♌ ☉ 6:23 / 19:28

2 ♋ M ♌ ☉ 6:24 / 19:26

3 ♋ ♌ TU ♌ ☉ 6:25 / 19:26

4 ♌ W ♌ ☉ 6:26 / 19:23

5 NEW ♌ TH ♌ ☉ 6:27 / 19:21

6 ♌ ♍ F ♌ ☉ 6:28 / 19:20

7 ♍ SA ♌ ☉ 6:29 / 19:18

8 ♍ SU ♌ ☉ 6:30 / 19:16

9 ♍ M ♌ ☉ 6:31 / 19:15

10 ♎ TU ♌ ☉ 6:32 / 19:13

11 ♎ W ♌ ☉ 6:33 / 19:11

12 ♏ TH ♌ ☉ 6:34 / 19:10

13 Set eggs to hatch on the Moon's increase, but not if a south wind blows.
♏ ♐ F ♌ ☉ 6:35 / 19:08

14 ♐ SA ♌ ☉ 6:36 / 19:06

15 The moon is perigee today, so sowing of crops should be avoided to prevent pest and fungus attack.
PG
♐ ♑ SU ♌ ☉ 6:37 / 19:05

16 1936: Emma Goldman helps write the English-language edition of the Confederación Nacional del Trabajo-Federación Anarquista Ibérica information bulletin, visits collectivized farms & factories, & travels to the Aragon front, Valencia, & Madrid.
♑ M ♌ ☉ 6:38 / 19:03

17 ♑ ♒ TU ♍ ☉ 6:39 / 19:01

18 ♒ ♓ W ♍ ☉ 6:40 / 19:00

19 Dig your horseradish in the full Moon for the best flavor.
FULL
♓ TH ♍ ☉ 6:41 / 18:58

20 ♓ F ♍ ☉ 6:42 / 18:56

21 ♓ ♈ SA ♍ ☉ 6:43 / 18:54

22 **Autumnal Equinox (Mabon)** The middle of harvest, a time of equal day and equal night, and for the moment nature is in balance. Reap what you have sown, finish up old projects and plans and planting the seeds for new enterprises or a change in lifestyle.
♈ SU ♍ ☉ 6:44 / 18:53

23 ♈ ♉ M ♍ ☉ 6:45 / 18:51

24 ♉ TU ♍ ☉ 6:46 / 18:48

25 ♉ W ♍ ☉ 6:47 / 18:48

26 ♉ ♊ TH ♍ MG ☉ 6:48 / 18:46

27 As the moon is in Apogee (at the farthest point of its elliptical orbit around the Earth) today, planting vegetable crops should be avoided.
♊ F ♍ ☉ 6:49 / 18:44

28 ♊ SA ♍ ☉ 6:50 / 18:43

29 ♊ ♋ SU ♍ ☉ 6:51 / 18:41

30 ♋ ♌ M ♍ ☉ 6:52 / 18:39

zodiac signs

Cancer Good for plant ing, irrigating, grafting + transplanting | Most Fruitful

Scorpio Exceptional for vine ripe growth | Fruitful

Pisces Good for low crops but a very poor time to make seedlings | Fruitful

Taurus A time to plant lettuce, cabbage, sturdy stalks - root crops | Semi-fruitful

Capricorn For planting potatoes + tubers | Semi-fruitful

Libra Time for seeding flowers | Semi-fruitful

Aries Time for cultivating + plowing where weed destruction is sought after. Chickens hatched now will be patient | Semi-barren

Sagittarius Best for destroying unwanted plant life | Semi-barren

Aquarius Suitable for planting of unwanted growth | Barren

Virgo Good for bringing order to the garden + general cleanup | Barren

Gemini Cultivation, hoeing + destroying unwanted growth | Barren

Leo Great for cultivating, hoeing + destroying unwanted growth | A time for killing weeds + unwanted growth. Cider should be drawn off for vinegar. A good time to cut hair - if curly hair is desired, cut hair in the increase of the moon.

RECIPE - Fall

by ZOE LATTA

take a whole head of garlic
cover it in honey
bake for 1 hour when its very hot
use knife to get garlic out

ALSO -

put 10 cloves of garlic in 2 cups of yogurt
serve with ground turkey, baked or stewed
tomatoes and some pasta

JER

THE NEW OLD TRADITIONS
THROUGHOUT THE YEAR

FALL EQUINOX: "The Birth of the Moon"
(On/around September 21st)

By nightfall we will once again celebrate a
birth, the birth of the Moon (ALL: Hooray!)
The Sun has finally made peace with its transition and has yielded to the power of its relative.
From now until the Spring Equinox the Moon
will begin to regain its influence over our days
and nights. Similar to the Spring Eq., the Fall
Equinox is a time to celebrate equality and re-
establish our connection to the seasons.

" Oh, mother, a bee has stung me," said a beautiful little
girl, as she came running in from the garden. " Never mind,
child,' replied the mother, ' it mistook thee for a flower."

Hive Minded:
The Past & Future Legacy
of The Honeybee

by SHAYNE GIPE

The keeping of honeybees has a richly storied history. The life of human agriculture has long been linked to that of bees for the crop pollination they perform and for the exquisitely sweet product of their labor. With the advent of urban rooftop beekeeping and with towns across the nation legalizing the installation of backyard hives, it is evident that a new era in the story of the honeybee, and thus the story of people, has already begun.

Though the exact beginning of beekeeping is impossible to determine, the virtues of the bee have been commented on by many figures throughout history, including Aristotle, Emily Dickinson, Francis Bacon, and Leonardo da Vinci. The honeybee and the community of its hive have come to serve as an archetype of industriousness, harmony, democracy, economy, and abundance. "For so work the honeybees," Shakespeare wrote, "creatures that by a rule in nature teach the act of order to a peopled kingdom." Each daily effort of the individual bee, including the colony's queen, is executed in service to the future of the hive. Because of this, humankind has constantly studied and identified with the honeybee, and looked to hive life as an exemplar of a healthy society.

J. H. MARTIN'S BEE-SUIT.

The recent upsurge of interest in the honeybee has been a bittersweet revival, however. The past several years have brought widespread media coverage of an epidemic that has been dubbed Colony Collapse Disorder. Heavy speculation still abounds regarding the reasons that entire colonies of bees vanish abruptly. Like so many of the environmental dilemmas we are challenged with today, much of the speculation links the outbreak to the damaging practices that human enterprise has inflicted upon the earth. As we have identified so closely with honeybees and rely on them heavily to pollinate and thus support our food system, any danger to them is a danger to us. "We could call it Colony Collapse disorder of the human being, too," biodynamic beekeeper Gunther Hauk warns in the documentary Queen of the Sun. Fortunately, the affinity that we continue to share with the honeybee has attracted many to answer this call for aid.

In doing so, experimental beekeepers are rethinking modern approaches to hive management. Such practices as migratory pollination services, and the use of plastic hive equipment and chemical pesticides, have been implemented in recent decades to combat the now widespread varroa mite and to yield greater honey harvests for mass-production. Like common large-scale agricultural conventions, also designed to maximize efficiency and optimize profit, these protocols are now under intense scrutiny. When working towards sustainable systems and vibrant communities, such ideologies fail to acknowledge the delicately interwoven relationships found in nature, particularly those that current science still fails to explain. Through our agricultural activity we have elected to play a role in the superintendence of the environment, which includes the wellbeing of the honeybee. We must find ways to nourish these connections in nature and among ourselves, as workers in our communities and as members of our collective human hive. In crafting stories of what we envision to be a healthy future, we may continue to take inspiration in this task from the honeybees, but must also offer our service to them in exchange.

"Men have forgotten this truth," says the fox to Antoine de Saint-Exupery's Little Prince. "You become responsible, forever, for what you have tamed."

The Little Prince asks in return: "What does that mean - 'tame'?"

"It is an act too often neglected," answers the fox. "It means to establish ties...if you tame me, then we shall need each other."

The world we now live in is the story of the many things that we have tamed. We have positioned ourselves as our world's co-authors and tied ourselves to every part of it. The earth is our inheritance, responsibility, and future. All aspects of our society, from agriculture and the keeping of bees to how we communicate and think, stem from what we as humans have domesticated and distilled from what once was wild. In doing so, we have fastened the future of the bee, along with much of the planet's life, to ourselves. The story we continue to tell does not belong to only us.

Moving ever-forward, into the winter and the countless seasons that lie beyond it, we may consider our many options in our roles as stewards here. The current shift towards sustainable and local thinking, and movement of young people back to the working of the land, suggest that quick and easy solutions are losing their glossy appeal. We are removing the chemical, plastic, and artificial from our hives and homes, and seemingly looking once more to the admirable honeybee for insight. We are returning to the same ideals the bees symbolize; community, co-operation, resiliency, and hard work are once again the orders of the day. This day and all the days to come are in our hands. Let us now work to wring sweetness from what our common future can hold.

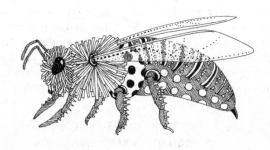

Farther Afield

by TOM WILLEY
Willey Family Farm, CA

Frosty temperatures of late have farmers of deciduous tree fruits hereabouts breathing easier, including almond growers. Local stone fruit Prunus cultivars require hundreds of winter chill hours (below 45° F) to remain fruitful. Only several weeks hence, almond buds will swell pink then riot into wedding veil-white bloom across three-quarter-million Central Valley acres. But cross-pollination nuptials must be attended by billions of honey bees, summoned from all over North America. Bakersfield friend, Joe Traynor is an Apis mellifera "travel agent" and "marriage broker", matching California almond orchardists with beekeepers across the nation since 1973. A former "Down on the Farm" radio guest, Joe has been sending me an excellent newsletter he pens for clients. Traynor's pre-bloom edition, warns 2013 will surpass recent years in the difficulty of providing strong bee colonies for pollination. The broker, esteemed by the apiarist community for his "good information and honesty", cites two primary reasons.

Supplies of beekeepers' most effective chemical treatment for debilitating Varroa mite parasites are exhausted after the sole foreign manufacturer ceased its production a year ago. Present in the US only since 1987, this vampire-like mite subsists on adult honeybee "blood", vectoring myriad viral diseases in the process. Secondly, last summer's drought conditions, affecting over 75% of the contiguous US (25% to a severe degree), resulted in drastically reduced bee forage that has left many regions' colonies in a "weakened nutritional state". Good bee pasture is further diminished as farmers continue to relentlessly pound a lucrative corn-soybean drum while abandoning traditional alfalfa, canola and clover throughout the Midwest.

Though it might appear otherwise, bees are not domesticated animals but rather a managed wild species. Apis health and immunity depends on a varied diet of pollen and nectar from numerous native plant species. Traynor's newsletter warns this one-two punch from Varroa mite and shrunken native habitat will catch ill-prepared orchardists, who couldn't stomach $150 contracts, scrambling for hives that may now rent for $200. Bees are well sustained nutritionally on almond bloom but need supplementation thereafter if colonies are to remain robust and healthy. Though a range war pitting seedless mandarin (Cuties) growers against apiarists has simmered down, Joe cautions traditional Valley citrus forage is mostly bee fast food as these blooms are nectar-laden but light on proteinaceous pollen. I've learned from Joe and top-notch local apiarists that, absent once-abundant natural resources, a beekeeper's skill as dietician and apothecary are assuming paramount importance.

FOR MORE INFORMATION GO TO
SCIENTIFICBEEKEEPING.COM

Keeping One Horse

To keep a horse, so that he may render the longest continued, and greatest amount of valuable service, it is essential that he should be provided with comfortable quarters.

The stable must be dry, warm, light, and with good ventilation; the latter to be managed on some plan that will not cause currents of air to pass over the inmate. A loose box is a great comfort to a horse. The term "box" seems to frighten most people. So far from being some complex arrangement, only within reach of the wealthy, it is as simple as possible. A reasonably wide stall may be used as one: but a few inches more width is desirable, depending on the size of the animal. If hard-worked, a horse rests more completely in a loose box than in a stall, and, when idle, he can exercise himself in it sufficiently to prevent stiffness or swelling of the legs.

A good bed keeps a horse clean when lying down, and aids his rest. Wheat straw is the best article of which to make it, and about five cwt. per annum will be needed. (1881)

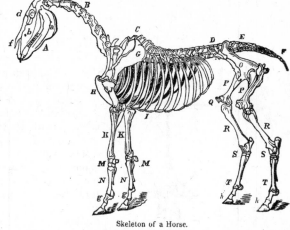

Skeleton of a Horse.

A. The Head.—a The posterior maxillary or under jaw.—b The superior maxillary or upper jaw.—c The orbit, or cavity containing the eye—d The nasal bones, or bones of the nose.—e The suture dividing the parietal bones below from the occipital bones above.—f The inferior maxillary bone.

B The Seven Cervical Vertebræ, or bones of the neck.—C The Eighteen Dorsal Vertebræ, or bones of the back.—D The Six Lumbar Vertebræ, or bones of the loins—E The Five Sacral Vertebræ, or bones of the haunch.—F The Caudal Vertebræ, or bones of the tail, generally about fifteen.— G The Scapula, or shoulder-blade.—H The Sternum, or fore-part of the chest.—I The Costæ or ribs, seven or eight articulating with the sternum, and called the *true ribs*; and ten or eleven united together by cartilage, called the *false ribs*.—J The Humerus, or upper bone of the arm.—K The Radius, or upper bone of the arm.—L The Ulna, or elbow. The point of the elbow is called the Olecranon.—M The Carpus, or knee, consisting of seven bones.

N The Metacarpal Bones. The larger metacarpal or cannon or shank in front, and the smaller metacarpal or splint bone behind.—g The fore pastern and foot, consisting of the Os Suffraginis, or the upper and larger pastern bone, with the sesamoid bones behind, articulating with the cannon and greater pastern, the Os Coronæ, or lesser pastern; the Os Pedis, or coffin-bone; and the Os Naviculare, or navicular, or shuttle-bone, not seen, and articulating with the smaller pastern and coffin bones.—h The corresponding bones of the hind-feet.

O The Haunch, consisting of three portions : the Ilium, the Ischium, and the Pubis.—P The Femur, or thigh.—Q The stiff joint with the Patella.—R The Tibia, or proper leg bone ; behind is a small bone called the fibula.—S The Tarsus, or hock, composed of six bones. The prominent part is the Os Calcis, or point of the Hock.—T The Metatarsals of the hind leg.

And a young colt could win anyone's heart.

CATTLE TIE.

Fig. 1.—*View from above.*

I GIVE YOU A DESIGN for a cheap, effective and simple cattle tie, which may be new to you or some of your readers. It is preferable to stanchions, as being more humane ; while it is always out of the cattle's way, so that they cannot possibly get it fastened anywhere so as to break it. It is easily put on and off.

The leather of which the straps are made should be made wet and *stretched* before using ; mine were not done so, and have become a little too long, which we correct by simply twisting one after the other one has been put on the horn.

I use them on my oxen, which stand two in the same stall, upon a raised platform, just long enough for them to stand on, with a gutter behind them to save manure.—J. R. G., in COUNTRY GENTLEMAN.

Fig. 139.

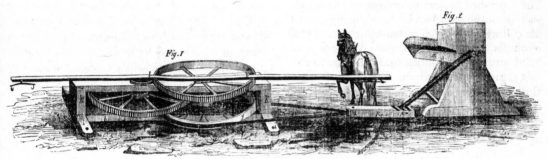

WARREN'S HORSE POWER AND THRESHING MACHINE.—(Fig. 50.)

EDITORIAL NOTE: Informal means of control-ling, with precision, the location of animals in the landscape or in the barnscape are no longer familiar to many new farmers. Laura Ingalls Wilder, in 'Little house on the Prairie" often talked about moving the stake for her family's dairy cow. Without fencing, or the means to fence, she wandered in a circle on the wide prai-rie. This continuum of fencing formality serves to remind us of the herding methods pre-fencing, and what human observation of animals, in far more intimate relationship with the land, might have yielded. Remember, barbed wire was in-vented only in 1865, electric fencing came into common practice only in the 1930's, whereas the Basques and French with their sheep, the Arabs with their camels, the Plains Indians with their untamed bison achieved their ends with far more modest tools. As we consider our pastures with NRCS maps and GIS pasture layouts on smart-phones, they day will soon come when we could 'wirelessly' herd our cattle in frequent intervals, with tremendous precision, and with only a 'hu-mane buzzing' into their ears. But if the ancient agronomy, or 'shepherds way' is absent from our study, then we won't know how to manage our animals, no matter how sharp the tool. What can we learn from the low-cost, high observation mode of animal management? - SvTF

The Fibershed Project

by REBECCA BURGESS

Another set of Sheep Fanatics can be found in Northern California (we cover 20 counties). The Fibershed project is looking to change the way we clothe ourselves by building a local textile culture in our region that utilizes local fiber resources to their full potential. We seek to rebuild processing facilities, and continue to connect designers and artisans with sheep farmers, alpaca ranchers, and any and all folks who are raising fiber and natural dyes in our region.

EDITORIAL NOTE:
Rebecca and her friends the shearers and shepherds have been the subject of an upcoming OURLAND.tv film produced by the Greenhorns. If you live in Northern California and work with fiber, or want to buy local fiber you can use their wonderful online marketplace, and connect with the larger community at her many events, workshops and trainings. - SvTF

MORE INFORMATION AT WWW.FIBERSHED.COM

FIG. 96. Holding a sheep and trimming its overgrown hoofs

Jim Barnett, Sheep Shearer: An Interview

as told to KAT SHIFFLER

Jim Barnett has worked as a traveling sheep-shearer for over 30 years. He answered some questions about wool markets, feeder lots, sheep-human relations, dental health and the good old days.

How'd you get started shearing sheep?

I started as a sheep-catcher for four or five guys in the Sioux Falls stockyards in South Dakota. It was the summer of 1971 and back then the shearers around this area didn't catch their own sheep. They had somebody else catch 'em and set 'em to 'em. And bag the wool; you know, a jump sacker. It's a six-foot bag and you have to hang it off the ground so you can put 200-300 pounds in it. You climb up the ladder and put the wool in the bag and then get inside and pack it down. You were always happy when it came to the top because then they would stop shearing and you could sew the bag shut.

So is that like the entry-level position? Sheep catching?

Yeah, its the ground floor if you want to get into the shearing business.

Back then I got paid by the head. It was a nickel apiece. I caught 1200 head in one day. But they were just tagging the rear ends. Otherwise it was 500-800 head a day – that's the average. They were only 70 pounds. Back then a fat lamb was 105 pounds. Now the the top market is buying 154 pounders right now.

So, what's the difference?

I'd say it's about 50 pounds.

But in general, how has the business changed since the 1970s?

Less sheep. A lot less sheep. Not nearly as many lamb-feeders around. They're getting to be bigger lots. Not so many mom and pops, where there'd be 500-1000 head on a place. They might still be a family operation when they're got 10-20,000 head around, but you know its just on a bigger scale anymore.

They've also made 'em bigger framed, and the packer wants them bigger because they're paying their help 'X' dollars an hour anyway. So they've made it so you've got a bigger cut of meat and sell more on the market to the consumer, you know. Like every other animal.

What's the secret for catching sheep?

Ha! You sneak up on the them! You just gotta walk slow and kinda see that they're in a trapped position and then you reach down and grab 'em by the jaw and nose. When I first started out they taught me to catch 'em by the leg and walk em backward and that works good but you've got to be pretty handy. Otherwise you can break a leg pretty easy. If you're right handed you grab the left side or the left rear leg and then just walk 'em backwards. When you get 'em to the sheerer you just flip 'em over – just give 'em a twist and they roll right into position.

Since then I've learned that you grab 'em by the nose and you walk 'em backward. I always call it the two-step. They seem to walk backwards and you just do the two-step. You can get 'em exactly where you want 'em, turn their head back toward their shoulder away from you and then they sit right down and kinda set themselves up. It's technique. It works great.

So, you travel around the Midwest sheering sheep. What's your job like on a daily basis?

You get up, eat breakfast, get to the job, set up your equipment, and they bring the sheep to you. You sheer 'em one at a time. Nowadays, I stick to Minnesota and Iowa mostly. But I travel to farms in Nebraska, Kansas, and Oklahoma.

With good help I can still sheer 80-100 head if they're good sheep.

What are you doing with the wool?

I'm buying for a company out of Illinois.They actually just broker it. They just split the seams and take it out of the truck, and then they go up conveyors and sort it all according to grade. They bag it again in 400-500 pound sacks, and then they ship it wherever it's supposed to go. A lot of it goes to China.

Okay, so these sheep, their byproduct is an international commodity. That seems like a good deal for you.

Yeah, that's one reason I took the job. Farmers have a product that they grow and they have to be sheared every year. And that's job security in my eyes. But the numbers have dropped quite a bit, as I say. I'm just not as willing to sniff around and find work a long ways away.

So, you work with bigger operations and feedlots. When is your busy season?

It'll pick up again in December. The feedlots if I can get back into them, then I'll be busy all the time. But the winter time is when they're going the hardest.

Don't they need that warm coat for the winter?

The lambs get muddy because we have our thaws – the January thaw and the February thaw – and so you want to take the wool off so they're not carrying 50 pounds of mud when they get sold. That's the biggest reason. The cleaner the wool, the better it cuts. And they do actually gain weight better once you take the wool off. They'll start eating a little more all the time.

It only takes ten days to get that seal back up where they can stand the weather again. Three days is a real critical time, but you're leaving a little bit of wool on them anyway. It doesn't take as long to get acclimated to the weather. Most people feed real good if they're shearing in the

wintertime. If you're not, you're going to have problems.

But when you shear is all in the preference in the grower. Most people sheer before they lamb. Personally I like to get 'em sheared 4-5 weeks before. Otherwise they're so pregnant they can't stand it.

So in these big operations, when are the lambs slaughtered?

Five to six months old. But that'd be someone who pushed 'em from day one. Your mountain lambs coming in that are grass and milk fed, they're good eating, but they want them to be bigger, so they put 'em on corn and protein for however long it takes.

Would you say you like sheep more than other animals?

Well, I guess I have to. But I really enjoy doing the chores when they're calm and they're used to you. A sheep knows a shepherd's voice and that's true. If you talk to them enough and give them a certain sound every time you feed them they realize who it is. Because you can go in the dark and they just run and bash into everything, but as soon as they realize you're there, they go, oh okay.

What does having sheep say about somebody? What kind of people have sheep?

Mostly they're pretty mild mannered, easy to talk to, easy to get along with. But I've seen people lose their temper to no end. If you're in a hurry with sheep, they're not going to do a thing for you. The calmer you are the better they are. They're the best motion detector on the farm, and I think they're one of the best emotion detectors too.

If somebody were to get into sheep today, what would you tell them?

Buy low, sell high. The best thing to do is buy young stock so you've got them around for a long time. But if you can't afford that buy some old ewes – 5 or 6 year olds from out west – and whatever you breed them to, save the ewe-lambs out of that. And go from there. You can buy old ewes for about $60 a head.

I've seen old ewes get up to 14 or 15. But at 5 or 6 years old you start seeing problems. It's hard to keep flesh on 'em after they lose their grinders or their molars in the back. They'll start losing them at about 5 or 6 years old. You call it a broken mouth. A lot of people look at the front teeth, but all those are for are picking grass, mowing grass. The back ones are what makes an animal survive. Because they've got to chew their cud.

Do you think your job, sheep shearing, is disappearing?

I would have to say it's shrinking, because the numbers are gone. It's probably good for a young person to do it as long as he has connections and is willing to travel. Because there's no way you're going to make a living now staying in one spot. That's just the bottom line.

In the 60's and 70's we used to go out to Gettysburg, SD and there were three or four of us who would sheer 125 head a piece and the rest 80-100. We'd be busy six days a week for five months within 25 miles of that town. And now I don't know if there's even 1,000 head of sheep around there all together.

We were living out of step vans. We'd just go down to the local park, and they had a shower we could use, and they wouldn't charge anything for it, and we didn't need electricity except to grind tools, and we'd just did that on the job site. Of course we'd sheer all day and shower up and go to the bar and grill and have our food and drink and tell stories.

So what happened to the sheep farmers in that area?

Well, corn and bean prices got high, and they realized they could raise that stuff in that country, because they were getting rain. So the more crops you can raise the more money you make, and you don't have all the chores. So the critters disappeared. It's a lot easier to get into the tractor with AC and heat and raise and lower that lever than it is to go out in 30-below and feed your critters and watch a baby lamb stick to the ground. You know. There's a lot of work. They call it animal husbandry for a reason. Basically you're married to them.

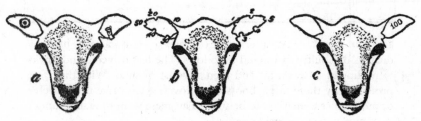

FIG. 97.   Three ways to ear mark sheep.   *a* 2 kinds of metal tag; *b* a notching system; *c* a number tattooed into the ear

Weaning Lambs.

Referring to this subject, W. H. Ladd of Jefferson co., O., (first rate authority on all sheep matters,) says :—My practice is to turn the lambs in with their mothers, after they have been separated some 12 hours, and as soon as they nurse, separate them again; then, after 24 hours, allow them to nurse once more. Since I have adopted this plan, I have never had an ewe's udder injured. Lambs should have a very little salt frequently, when first weaned, as the herbage lacks the large proportion of salt which the mother's milk contains. But great care should be used not to give them much salt at once, or it will set them to purging; and if a lamb commences to purge soon after being taken from the mother, it seldom ever recovers from it.

THE ILLUSTRATED ANNUAL REGISTER OF RURAL AFFAIRS
AND CULTIVATOR ALMANAC FOR THE YEAR 1862,
THOMAS, J.J. (JOHN JACOB)
WWW.ARCHIVE.ORG

TAR FOR SHEEP.

A gentleman who keeps a large flock of sheep, says that during the season of grazing he gives his sheep *tar*, at the rate of a gill a day to every twenty sheep. He puts the tar in troughs, sprinkles a little fine salt over it, and the sheep consume it with eagerness. This preserves them from worms in the head, promotes their general health, and is thought to be a specific against the rot.

EDITORIAL NOTE: An alternative to cryovac - tar was also used by drovers of poultry in Ireland as follows. The feet of geese and ducks were dipped in warm tar and then sawdust or sand. With their feet protected by these tarry 'boots' the drover could drive them twelve or more miles to market, to be sold. The geese went to market under their own power. - SvTF

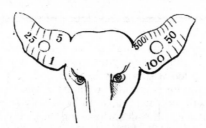

Registering Sheep.

The above cut illustrates the German mode of ear-marking and regularly numbering the sheep belonging to a flock, so that each individual can be distinctly registered:

Each slit in the lower rim of the right ear represents.......   1

Each slit in the upper rim of the right ear represents......................   5

Each slit in the lower rim of the left ear represents 100

Each slit in the upper rim of the left ear represents 500

The central hole in the right ear represents..............................   25

The central hole in the left ear represents................................   50

In the above figure :

7 slits in the upper rim of the left ear, 500 each 3,500

4 slits in the lower rim of the left ear, 100 each   400

The central hole in the left ear   50

4 slits in the lower rim of the right ear, 5 each   20

4 slits in the lower rim of the right ear, 1 each   4

The central hole in right ear...   25

Number of the sheep.......... .. 3,999

Hog-sty.—This building should contain one apartment in which the swine shall be perfectly dry, and
well provided with straw for their bed.   Another part
of it should be open, and without a wooden floor; for
swine will not well bear to be wholly secluded from
the weather.   Besides, this open part may be extended, so as to afford a fine place for making manure.
It should be lowest in the middle, and no water should
ever run from it.   With proper care, many loads of
valuable manure will be made every year, where these
animals are kept.   The trough should be made fast at
the upper side; and if the edges are covered with iron,
it will be well.   There should be stakes before it, so
thick that one swine only can get his head between
any two of them.   The keeping of swine is very profitable to the farmer.   Indian corn is the best food for
them, but it should be ground, or boiled till it is soft.
Like human creatures, however, they require some
variety in their food.   Steamed potatoes may be given
them with much advantage.

EDITORIAL NOTE: According to regulations, in the United States you are not allowed to feed restaurant food-scraps to pigs without cooking them first. This is to prevent the passage of trichinosis (a worm) from pork scraps back into the living pork population. Pigs are allowed to eat spoiled yogurt, and other grocery items pre-consumer-- but not post consumer. (See Regulation Number 74 FR 15215-15218 for more information.) Many of us feed spoiled vegetables, sour milk, brewers grains, mangal roots and undesirable farm residues to our pigs-- all in an effort at gain with less feed costs. Pigs can be rotated through after corn has been threshed, or after vegetables are finished for the year-- we do not hesitate to turn them loose on waspy apples and leftover ice-cream, moldy bread from the bakery, or dumpstered produce in an urban context. But are we considering their pigly needs, cost per gain, caloric availability roasted or raw, crushed or cracked, in which ratios with protein and minerals? Usually no, usually we leave this to the feed manufacturers and blenders of commodities and mineral supplement who either bag for us, or deliver it ready-mixed to our small silos. Until the 1930's even, the majority of farmers mixed their own rations, and usually grew their feedstuffs themselves. As animals come back onto our vegetable operations, so too, slowly, can we anticipate the re-entry of farm-grown feed, processing, pelleting, mixing, roasting and a fine business selling chick-scratch to homesteaders. - SvTF

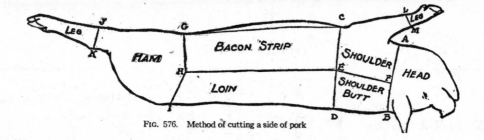

FIG. 576. Method of cutting a side of pork

EXPERIMENT IN COOKING FEED FOR HOGS.

A Wayne County farmer has accurately tested the results of cooking feed for swine, and presents the following figures :

The experiment referred to was conducted with two pens of hogs, which were carefully weighed, the gains noted, and the food in each case also weighed or measured. The hogs selected for the experiment were all grade Chesters, and, with one exception, nearly of the same age, weight, condition, &c. Pen No. 1 contained three hogs, whose live weight was nearly 1,000 pounds. They were fed all the corn they would eat up clean —the three consuming forty-five pounds of corn daily. After being fed seven days, they were again weighed, when it was found that they had gained ten pounds each. By calculation we find that during the seven days this pen of hogs consumed five bushels and eight quarts of corn, costing $6.66. The gain being thirty pounds, we see that thirty pounds of pork cost $6.66, and would have sold at the time for $2.55. Pen No. 2 contained two hogs, one of which weighed alive six hundred pounds, and the other nearly four hundred pounds. They were fed all the cooked meal they would eat—the two consuming twenty-five pounds of meal per day. The respective gains of each were five and seven pounds, the smaller hog gaining five pounds per day, and the larger seven pounds.

By calculation we find that the pork made from whole-corn cost a trifle over 22 cents per pound, while that made from cooked meal cost 4½ cents per pound.

I am aware that seven pounds may seem to some to be an extravagant gain for a hog in one day, but it must be remembered that this was a very large hog, and the experiment conducted in very favorable weather. I succeeded in increasing the gross weight of this hog five pounds each day for weeks together in rather unfavorable weather. I believe we can make hogs profitable, if we feed a few all they can eat, but when we undertake to keep fifteen hogs upon the feed of five, we must expect disappointment and loss. It is also equally essential to have *a good breed* of hogs to feed.

THE ILLUSTRATED ANNUAL REGISTER OF RURAL AFFAIRS
AND CULTIVATOR ALMANAC FOR THE YEAR 1869,
THOMAS, J.J. (JOHN JACOB) WWW.ARCHIVE.ORG

Gestation of Animals.

KINDS OF ANIMALS.	Proper age for Reproduction.	Period of the Power of Reproduction.	Number of Females for one Male.	Period of Gestation and Incubation.		
				Shortest Period.	Mean Period.	Longest Period.
		YEARS.		DAYS.	DAYS.	DAYS.
Mare,...................	4 years.	10 to 12	....	322	347	419
Stallion,.................	5 years.	12 to 15	20 to 30			
Cow,....................	3 years.	10	....	240	283	321
Bull,...................	3 years.	5	30 to 40			
Ewe,...................	2 years.	6	....	146	154	161
Tup,...................	2 years.	7	40 to 50			
Sow,...................	1 year.	6	....	109	115	143
Boar,...................	1 year.	6	6 to 10			
She-Goat,...............	2 years.	6	....	150	156	163
He-Goat,...............	2 years.	5	20 to 40			
She-Ass,................	4 years.	10 to 12	....	365	380	391
He-Ass,................	5 years.	12 to 15				
She-Buffalo,............	....	....	....	281	308	335
Bitch,.................	2 years.	8 to 9		55	60	63
Dog,..................	2 years.	8 to 9				
She-Cat,...............	1 year.	5 to 6	....	48	50	56
He-Cat,...............	1 year.	9 to 10	5 to 6			
Doe-Rabbit,.............	6 months.	5 to 6	....	20	28	35
Buck-Rabbit,............	6 months.	5 to 6	30			
Cock,.................	6 months.	5 to 6	12 to 15			
Turkey,................	....	....	....	24	26	30
Hen,..................	....	3 to 5	....	19	21	24
Duck,.................				28	30	32
Goose,................				27	30	33
Pigeon,................				16	18	20

THE "RED" BERKSHIRE SOW "BELLE."—*Engraved for the American Agriculturist.*

"The production of food should therefore be assisted. The burdens of agriculture should be alleviated. The small cultivator was to be encouraged and held in honour; the idle consumer viewed with reprobation. Luxury he defined as the abuse of wealth. An unequal distribution of wealth is prejudicial to production, for the very rich are "like pikes in a pond" who devour their smaller neighbours. Great landowners should live upon their estates and stimulate their development,—not lead an absentee life of pleasure in the metropolis. Interest should be reduced, public debts extinguished, and a ministry of agriculture created to bring to agriculture the succour of applied science, to facilitate the development of canals,communications, drainage, and so forth.

The state is a tree, agriculture its roots, population its trunk, arts and commerce its leaves. From the roots come the vivifying sap drawn up by multitudinous fibres from the soil. The leaves, the most brilliant part of the tree, are the least enduring. A storm may destroy them. But the sap will soon renew them if the roots maintain their vigour. If, however, some unfriendly insect attack the roots, then in vain do we wait for the sun and the dew to reanimate the withered trunk. To the roots must the remedy go, to let them expand and recover. If not, the tree will perish."

VICTOR DE RIQUETI, MARQUIS DE MIRABEAU
FROM PRE-REVOLUTIONARY FRANCE 1775

A Little About
The Ancona Duck

by EVAN GREGOIRE
OF BOONDOCKERS FARM, OREGON

The Ancona duck is a special dual purpose breed that was found in Great Brittan and brought to this country by Dave Holderread. They have been raised in the United States for several decades and were exhibited in 1983 in Oregon. The Ancona averages around 6 pounds and is a quite larger than its close relative, the Magpie duck. It has a medium sized oval head and a bill that is slightly concave along the top ridge. The neck should arch forward slightly and the body carriage is approximately 30 degrees above horizontal, a Runner trait. The broken, non-symmetrical feathering is unique - there is no set plumage pattern with the unique Ancona.

"The Ancona is a hardy, adaptable, all-purpose duck. It is an excellent layer, typically laying 210-280 white, cream, or blue eggs yearly. The Ancona also grows relatively quickly, and produces high quality meat that is more flavorful and less fatty than that of most Pekin ducks."

-Dave Holderread

The Ancona gives chefs, bakers and foodies a world of new culinary opportunities to experiment with. The meat is wonderfully dark and rich, and if you have even a small breeding program there are always a few extra males around for the table. Keeping them healthy is important, and keeping the forage fresh is vital to that. If rotated on pasture with a combination of hedgerows, long grass, and short grass, Anconas can forage for amazing amounts of their diet. When letting the ducks forage the only trick is getting the area fenced so they can be kept away from crops, gardens, and flower beds. They tend to stay where you fence them and herd easily if they get out. None of the laying breeds of ducks will fly very much or very high. Fencing the Anconas with a two-foot fence framed out with chicken wire or bamboo stakes work wonders, though a fence this short will not stop a scared duck or keep a duck separate from friends on the other side. To keep larger predators away from the ducks, you'll need perimeter fencing strong enough to exclude dogs and coyotes. Four-foot electric poultry works great. The night pen or house also needs to exclude night predators. Poultry wire isn't useful in constructing night pens, since coyotes, foxes, and raccoons can all tear through it - hardware cloth, though more expensive, is better. Protecting the birds can also be done successfully

This Airplane View Shows Poultry Tribune Experimental Farm Where Staff Members Keep Constantly in Touch With Poultry Problems

Feeding the barnyard fowl was, often enough, a boy's first responsibility.

with the help of livestock guardian dogs.

The American Livestock Breeds Conservancy's 2000 census of domestic waterfowl in North America found only 128 breeding Ancona. While four people reported breeding them, only one primary breeding flock with 50 or more breeding birds exists, located at Boondockers Farm in Oregon. Boondockers Farm's number one mission is breeding the Ancona Duck and furthering the hardy genetics. Boondockers acquired all of Dave Holderread's Ancona stock in 2010, to continue his 30+ years of work to raise the most productive Ancona ducks possible while keeping all breed traits intact. There is a critical need for more conservation breeders of Ancona ducks. Their excellent laying ability, tasty meat, and calm dispositions make them a great addition to any small farmstead or backyard producer's flock. Even small breeding flocks can help keep the genetic diversity going, so don't be hesitant to start small - everything helps.

EGG·A·TORIALS

Better breeding, good management and efficient marketing
are the basis of permanent prosperity in the poultry business

⟡ ⟡ ⟡

Yes, someone will, no doubt, say, "you favor commercial poultry feeds because these companies advertise in Poultry Tribune." True enough, they do advertise, but it is our honest conviction, based on our own experience, that ready mixed feeds are going to play an ever increasing part in the poultry industry. They have justified themselves. They are economically sound.*

We suggest that the farmer use for scratch feed the corn and oats and what other small grains he grows, and that he leave his mash mixture in the hands of experts who know exactly how to mix a ration that will produce eggs at the lowest cost.

Subscribers who have opinions may wish to write to us.

August, 1930 — POULTRY TRIBUNE — Page 7

POULTRY HOUSE.—Every farmer should have a good, convenient poultry house, properly constructed, sufficiently large to contain the number of birds he desires. It should be warm and dry in the winter, well ventilated and kept scrupulously clean. The house should not be over crowded, but just large enough. Nothing is made by over-crowding the hennery ; on the contrary it will prove detrimental. The fowls must be fed regularly and at stated periods. They must have plenty of pure water at hand at all times—this is of as much importance to the health of the brood as proper food. If possible, they should also be given, in addition, a plat of grass for a run. Place within the hennery a dust heap ; this may consist of wood or coal ashes, sand, or dust from the streets. It should be kept under a cover, so that it will not become drenched with rain or snow, and to it the fowls should have access at all times, to dust and thereby rid themselves, in a great degree, of the numerous parasites which infest them.—*Poultry Standard.*

"An egg a day keeps the hatchet away!"

WHEN you reach in the nests on one of those September mornings...and discover a pullet egg ...the very first one...someone is just bound to hear about it soon! No doubt that egg is small... but how important! Important enough to be news to all the neighborhood.

It means more than just another egg. It means your pullets are starting to lay just as egg prices are starting to go up...as they always do in the fall. But to *keep* them laying...that's the big job. That's the job you can tend to *now!* Just by feeding Purina Growena Chow (mash) and Purina Intermediate Hen Chow (scratch) through these summer months.

Together these two feeds contain every single thing your pullets need to build themselves into birds that will lay at 16 to 20 weeks of age...every single thing they need to grow strong enough and big enough to *keep* laying through October... November...December...January...February ...with never a stop! These are the months eggs are worth money...these are the months for you to make money. These Purina Poultry Chows before your pullets *now* will do that very job for you!

 AT THE STORE WITH THE CHECKERBOARD SIGN

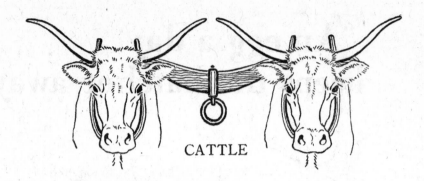

CATTLE

Manure of Different Classes of Animals.

Cows in full milk or with calf, secrete from their feed, great quantities of valuable substances which fattening cows or oxen will not withdraw from what they eat—hence the manure of the milch cows is not worth nearly so much as that of fattening animals. This must be evident from the obvious fact, that out of the milk, or what would be milk, the entire structure of a 5-weeks-old calf is formed. The calf continues to grow and learns to eat the same food that the cow does, and for several years is building up his frame of heavy bones; all the valuable ingredients permanently entering into his system, of course come out of his feed, and would, were he a full-grown steer, have passed into the manure. Many farmers are likely to undervalue the important differences in the quality of the excrements of different classes of the same kind of stock.—The differences which we have alluded to, of course exist as well in the manure of other kinds of animals as in that of neat cattle. Let us then bear in mind that keeping milk-giving and growing animals is a great tax upon the land, that fattening animals make rich manure heaps, and that full-grown male animals draw much less upon the soil than females bearing young and giving milk.

Milk Products

Cream is simply very rich milk, in which the proportion of fat has been increased by the removal of part of the liquid contents, either by setting and skimming or by the action of a separator (p. 454). It may contain from 15 to as much as 60 per cent of fat, depending on how thoroughly the skimming or separating is done. *Ice cream* is the result of stirring or beating sweet cream (or a cooked custard of cream, milk, eggs, etc.) carrying some flavoring material, in a vessel surrounded with crushed ice or brine. As far as the milk or cream is concerned, the composition and food value is exactly the same as in the unfrozen product. Some ice cream manufacturers mix sweet butter with skimmilk and force it through a special machine to distribute the fat. Thus they save freight on the moisture in fresh cream and lessen the chance of its spoiling.

FIG. 493. Percentages of materials in average cream

FIG. 494. Percentages of materials in separator skimmilk

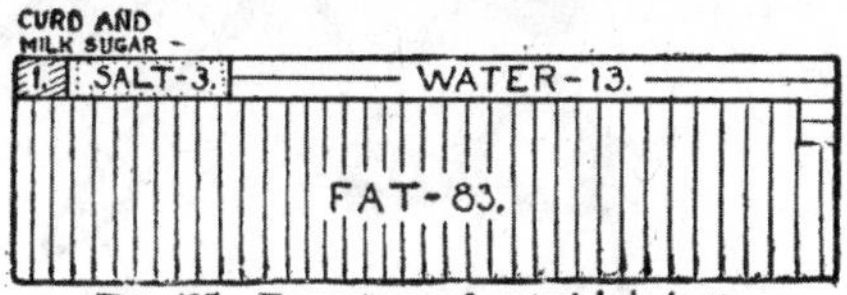

FIG. 495. Percentages of materials in butter

Milk of average quality will produce, per 1,000 pounds, about 145 pounds of cream and 850 pounds of skimmilk (a loss of 5 pounds in the form of slime, etc. occurring in the separator). The 145 pounds of cream will make about 42 pounds of butter, leaving 100 pounds of buttermilk, 3 pounds being lost in the process.

Skimmilk is the milk serum with what little fat is left after the skimming or separating operation. Since it is just as rich in casein and albumen as whole milk, and since these materials are especially needed by growing animals, it has a food value that is not as fully appreciated as it should be.

Butter (Chapter 43) consists of the fat particles collected and worked into a compact mass from which most of the liquid and the other milk solids have been removed. It should contain at least 80 per cent fat, the rest being water, and, roughly, 1 per cent curd and milk sugar, and 3 per cent salt.

FIG. 496. Percentages of materials in buttermilk

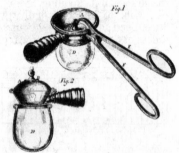

PARASOL FOR PLOW-BOYS. *Nothing in the record indicates that this patented umbrella ever became popular; but it does prove that inventors didn't overlook the poor husbandman.*

Almost a century ago, the august Scientific American displayed this milking machine—Calvin's Patent Breast Pump—as a wonderful example of the fruits of Yankee ingenuity. Old farmers agree that hand milking probably drove more men off the farm than any other task. This device was expected to stop that trend. It didn't.

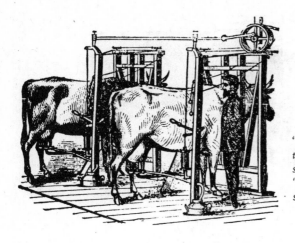

"It Imitates the Calf..." That's what the *Hydraulic Cow-Milking Machine Co.* said in 1868 of its patented rig which would "fit any cow" and was "simple, durable and self-adjusting." But the ideal mechanical milker did not appear until the 1920's.

FARM HACK

A COMMUNITY FOR FARM INNOVATION
AN OPENSOURCE TOOL DESIGN PROJECT

2013 Events In Minnesota, Michigan, California,
New Hampshire, Vermont, New York
join the network, build adaptable, appropriate
tools for your sustainable farm.

FOR MORE INFORMATION:
farmhack.net

table and jar structures preserve
the cavernous aromas of transition
pickle & pumpkin, the harvest seeds
collective return and furry reverie

–Leonora Zoninsein

OCTOBER

———

FLURRIES

OCTOBER 2013

zodiac signs

Cancer — Good for planting, irrigating, grafting, transplanting. *Most Fruitful*	**Scorpio** — Fruitful.	**Pisces** — Good for root crops but is a very poor time to make seed-free.
Taurus — A time to plant lettuce, cabbage, sturdy stalks + root crops. *Semi-fruitful*	**Capricorn** — Good time to plant tubers + root crops. *Semi-fruitful*	**Libra** — Time for seeding of hay, grass, flowers and crops. Most frequently used. *Semi-fruitful*
Aries — Time for cultivating + plowing where weed destruction is sought after. Chickens hatched now will be patient. *Semi-barren*	**Sagittarius** — Best for destroying unwanted plant life. *Semi-barren*	**Aquarius** — Suitable for seeding + destroying of unwanted growth. *Barren*
Virgo — Used for bringing color to the garden + general cleanup.	**Gemini** — Good for culture, digging + destroying growth.	**Leo** — A time for killing weeds + unwanted growth. Cultivate the flowers off its vintage. A good time to cut hair - if curly hair is desired, cut hair in the increase of the moon.

---

8 — In 1982, all labor organizations in Poland, including Solidarity, were banned.
♏ TU ♍ ☉ 7:00 18:26

16
♓ W ♍ ☉ 7:09 18:14

24 — In 1945, the United Nations officially came into existence as its charter took effect.
♊ TH ♍ ☉ 7:17 18:02

1 — The time it takes for the Sun, Moon and planets to make a complete circuit around the celestial sphere as viewed from the Earth is called the sidereal period, from the latin word "sider" for star. The Sun's sidereal period is one year, while the Moon's is 27 days.
♌ TU ♍ ☉ 6:53 18:38

9
♏ W ♍ ☉ 7:01 18:25

17
♓ TH ♍ ☉ 7:10 18:12

25 — AG
♊ F ♍ ☉ 7:19 18:01

2
♌ W ♍ ☉ 6:54 18:36

10 — PG
♏ ♐ TH ♍ ☉ 7:02 18:23

18 — The penumbral eclipse of the moon may be seen in the Eastern sky as the moon is rising today. Avoid planting until midday tomorrow.
FULL ♓ F ♍ ☉ 7:11 18:11

26
♊ ♋ SA ♍ ☉ 7:20 18:00

3
♌ ♍ TH ♍ ☉ 6:55 18:34

11
♐ F ♍ ☉ 7:03 18:22

19
♓ ♈ SA ♍ ☉ 7:12 18:09

27
♋ SU ♍ ☉ 7:21 17:58

4
NEW ♍ F ♍ ☉ 6:56 18:33

12 — Harvest Festival celebrated in Nigeria
♐ ♑ SA ♍ ☉ 7:04 18:20

20
♈ SU ♍ ☉ 7:13 18:06

28
♌ M ♍ ☉ 7:22 17:57

5
♍ SA ♍ ☉ 6:57 18:31

13
♑ SU ♍ ☉ 7:05 18:18

21 — Revolution Day: Commemorates the bloodless revolution of 1969 in Somalia
♉ M ♍ ☉ 7:14 18:07

29
♌ TU ♍ ☉ 7:23 17:56

6
♍ ♎ SU ♍ ☉ 6:58 18:30

14
♑ ♒ M ♍ ☉ 7:15 18:17

22
♉ TU ♍ ☉ 7:15 18:05

30
♌ ♍ W ♍ ☉ 7:24 17:55

7
♎ M ♍ ☉ 6:59 18:28

15
♒ Tu ♍ ☉ 7:07 18:15

23
♉ W ♍ ☉ 7:16 18:04

31 — 1925, Australia: Radio station 2KY begins broadcasting in Sydney, NSW; it later becomes the world's first labor-owned radio station.
♍ TH ♍ ☉ 7:25 17:53

Fig. 196.—*Ventilation for Warm-Air Furnace.*

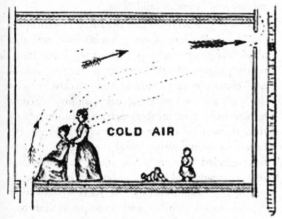

COLD AIR

Fig. 197.—*Unequal Ventilation.*

THE NEW OLD TRADITIONS
THROUGHOUT THE YEAR

HALLOWEEN: "Long Live the Dead"
(On/around October 31st)

As we near the Sun's death our attention draws toward taboo subjects. We celebrate the living by familiarizing ourselves with what has passed, in a tradition that is expressed in a myriad of ways by many different cultures united by a recognition that the veil separating the living from the dead is perhaps thinnest at the time when October leans into November. Whether it be expressed as Samhain, a Celtic holiday celebrating the final harvest of the year, or through the tradition of trick-or-treating, where the young rule the old, Halloween is an opportunity to embrace the inversion of life's assumed norms. It is also a time to note the impermanence of physical life by honoring those whose bodies have returned to the earth, a practice many people before us have thought it right to do.

The Coat

by JESSICA ELSAESSER

It's true, what they say,
there are two creatures
on your shoulders,
weightless and irascible.
Sometimes it seems
they even argue
in your sleep,
waking you will find
your mind's made up
already, a coat you carry,
that makes you spurn
the weather or
the burden.

Land Rant

by ANTONIO ROMAN-ALCALA

When Richard Heinberg says, "we need 50 Million new farmers to create a more sustainable food system", I can't help but agree. Yet I also wonder, where are those farmers to farm? On whose land? This is an especially pertinent question for new farmers like me who live in areas with extremely high agricultural land values (according to the USDA, the most expensive prices are found in the Northeast, California, Florida, and Virginia).

Unless you are already wealthy, or extremely lucky (or hooked into your local Farmlink chapter!), securing land for your new farm is difficult and will only become more so as more and more new farmers enter the search for land. But we shouldn't be competing for scarce resources; we should be joining forces as the new farmer movement to reconstruct access to land.

The young farmer revival in the U.S. seems to have an unconventional perspective on what constitutes social value: like many before us, we prioritize values beyond money itself, and feel this alternative sense of value – prioritizing equality, responsibility, sustainability – must permeate our relationships with the land that sustains us. In contrast, the conventional measure of value as purely monetary has led to many social ills, including the ruining of many pieces of land. And so many of my generation question high levels of money-making as a goal for a particular piece of land, just as we question endless economic growth as a goal for society. We aren't against economies; we want to replicate meaningful, challenging, yet joyful employment opportunities. Yet we are beginning to see endless growth more correctly as a problem rather than as the panacea politicians make it out to be.

Private property rights have a problematic history: we can see benefits of private property (in contrast to other attempted forms of property relations), yet private property rights have allowed the continued destruction of the planet and the subjugation of a large percentage of the population to the destructive whims of property-owning elites. Further, the right to property has reached a beyond-reproach, almost mythological standing in our ultra-individualist, ultra-capitalist ideological milieu. This won't change overnight, but it needs to change. And if we are to go about changing the role of property in society, it is better for it to happen slowly and deliberately rather than suddenly (no one is looking for a violent change up in ownership, French Revolution-style).

So what if we could "kill two birds with one stone"? What if we could achieve greater access for young farmers to farmland, while tempering some of the worst ills brought on our country by misuse and abuse of land?

Article 5, XXIII of the Brazilian constitution mandates that land serve a "social function" and allows the government to "expropriate [rural land] for the purpose of agrarian reform." Similar "social use" clauses in our state constitutions could open up legal avenues to make cases against current agribusiness and mineral extraction interests in their decimation of our country's natural resource base. Simultaneously, socially and environmentally-minded farmers could make the case that they are ready and willing to put this land to truly productive use, building the food system that creates social stability and widespread benefits, gustatory, ecological, and economic among them.

Thankfully, we wouldn't be starting from scratch, as there are already legal instruments for the expropriation of privately held property for public uses. Though its history is certainly fraught with abuse by the state, largely in favor of politically powerful interests, "eminent domain" precedents and the "Takings" clause of the 5th amendment do allow for the idea that our government may intervene in private property markets to achieve socially beneficial ends.

The challenge for us is to define those ends well and maneuver the legal system to place the onus of land use righteousness on the current land-owning elite. In this project, un-used and mis-used parcels of land become contested spaces for re-narrating the story of Anglo America; do we continue our trajectory of consolidation, exploitation, degradation, and alienation? Or do we allow a new generation of dedicated and diverse, interconnected and interacting, locally-focused yet globally-minded food producers to re-make our nation's landscape?

With legal expertise, risky and radical challenges to private property, and long term effort, new farmers of this land mass may have a chance to remake the landscape, and with it our collective future.

A typical castle in Germany, surrounded by its village.
This is the feudal agrarian state, painted in a romantic manner
for tourists such as Lord Byron.

On The Building of Farm Houses

by SALLY MCMURRY

*The Cultivator, 1846, From Families & Farmhouses
In 19th Century America*

My better half, sitting at my elbow, says she would like to have a plan of a house or cottage, suitable for a large family, in which all the rooms should be on one floor... for, she says that running up and down stairs makes the woman look old when they are young, and that a cellar kitchen is an abomination. And further, she thinks that what little scolding and fretting is heard among them, is owing very much to the ill judged plans of their houses.

By 1859 " Moore's Rural New Yorker" described a cooperative process of design: "Usually among our farming community the plan of a house is got up by the builder, or, rather, the proprietor decides upon the size and general shape, and then the carpenter subdivides it into apartments under the direction of the female head of the family."

The culture of progressive agriculture, then, had a direct bearing upon the design process, since progressive farm families valued an experimental and cooperative approach to design. This approach in turn influenced the actual forms that the plans assumed.

1846 IN THE CULTIVATOR FROM FAMILIES & FARMHOUSES IN 19TH CENTURY AMERICA

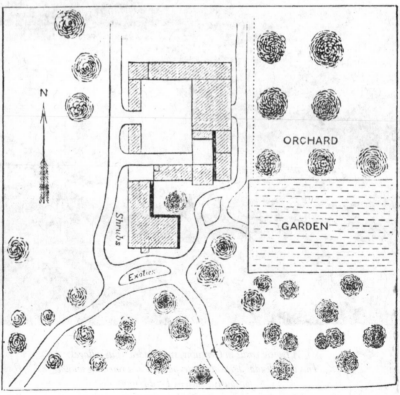

Plan of Buildings and Grounds.—(Fig. 48.)

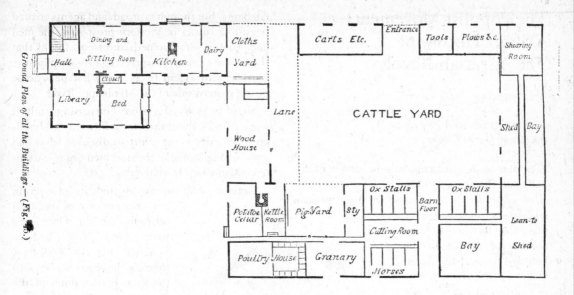

Ground Plan of all the Buildings.—(Fig. 50.)

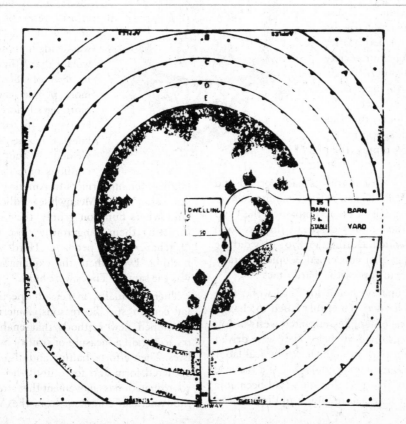

The Populist Movement: A Short History of The Agrarian Revolt

by LAWRENCE GOODWYN

*Describing the movement of the 1870s
that gave birth to the Grange*

"The passionate events that are the subject of this book had their origins in the social circumstances of a hundred years ago when the American population contained huge masses of farmers. A large number of people in the United States discovered that the economic premises of their society were working against them. These premises were reputed to be democratic - America after all was a democratic society in the eyes of most of its own citizens and in the eyes of the world-but farmers by the millions found that this claim was not supported by the events governing their lives.

The nation's agriculturalists had worried and grumbled about ' the new rules of commerce' ever since the prosperity that accompanied the Civil War had turned into widespread distress soon after the war ended. During the 1870's they did the kinds of things that concerned people generally do in an effort to cope with " hard times".

In an occupation noted for hard work, they worked even harder. When this failed to change things millions of families migrated westward in an effort to enlist nature's help. They were driven by the thought that through sheer physical labor they might wring more production from the new virgin lands of the West than they had been able to do in their native stats of Ohio and Virginia and

Alabama. But though railroad land agents created beguiling stories of Western prosperity, the men and women who listened, and went, found that the laws of commerce worked against them just as much in Kansas and Texas as they had back home on the eastern side of the Mississippi River.

So in the 1870's, the farmers increasingly talked to each other about their troubles and read books on economics in an effort to discover what had gone wrong. Some of them formed organizations of economic-self help like the

Grange, and others assisted in pioneering new institutions of political self-help like the Greenback party (and Non Partisan League). But as the hard times of the 1870's turned into the even harder times of the 1880's, it was clear that these efforts were not really going anywhere. Indeed, by1888 it was evident that things were worse than they had been in 1878 or 1868. More and more people saw their farm mortgages foreclosed. As everyone in rural America knew, this statistic inexorably yielded another, more ominous one: the number of landless tenant farmers in America rose steadily year after year. Meanwhile millions of small landowners hung on grimly, their unpaid debts thrusting them dangerously close to the brink of tenantry and peonage. Hard work availed nothing. Everywhere the explanation of events was the same: " Times were hard."

Then gradually, in certain specific ways and for certain specific reasons, American farmers developed new methods that enabled them to try to regain a measure of control over their own lives. Their efforts, halting and disjointed at first, gathered form and force until they grew into a coordinated mass movement that stretched across the American continent from the Atlantic coast to the Pacific. Millions of people came to believe fervently that a wholesale overhauling of their society was going to happen in their lifetimes. A democratic " new day" was coming to America. This whirlwind of effort, and the massive upsurge of democratic hopes that accompanied it, has come to be known as the Populist Revolt. This book is about that moment of historical time. It seeks to trace the planting, growth and death of the mass democratic movement known as Populism. "

Grange Halls in the United States

OBAMA'S VEGETABLE GARDEN *by* BECCA BARNET

nasturtium, marigold, zinnias

rhubarb

red leaf lettuce

shell peas

broccoli

fennel

onions

sugar snap peas

red romaine lettuce

dill, cilantro, parsley

butterhead lettuce

kale, collards

blueberries, raspberries, blackberries

green oakleaf lettuce

carrots

shallot

onions

spinach

The Husbandry School

What is husbandry? We are not talking about animal husbandry (the care and management of animals). Or crops. Or even husbands. We are talking about something VERY much more. Husbandry is an ancient word which means nothing less than 'the care and management of nature and resources for living'. This meaning has all but gone into obscurity. We are going to help renew that meaning. Husbandry is an ancient art which we believe will play a vital role inthe transition we all must make towards a sustainable future.

WWW.HUSBANDRY.CO.UK

La Chasseur Francais - The French Hunter Great Pyrenees Guard Dog

"But this was not going to be just any garden. Here even the tomatoes and beans would have a view of the towering Washington Monument. But one thing I knew was that I wanted this garden to be more than just a plot of land. I wanted it to be the starting point for something bigger. I was alarmed by reports of skyrocketing childhood obesity rates and the dire consequences for our children's health. And I hoped this garden would help begin a conversation about the food we eat, the lives we lead, and how all of that affects our children. I also knew that I wanted this White House garden to be a "learning garden", a place where people could have a hands-on experience of working the soil, and children who have never seen a plant sprout could put down seeds and seedlings that would take root." —Michelle Obama

THE NITROGEN CYCLE *by* BROOKE BUDNER

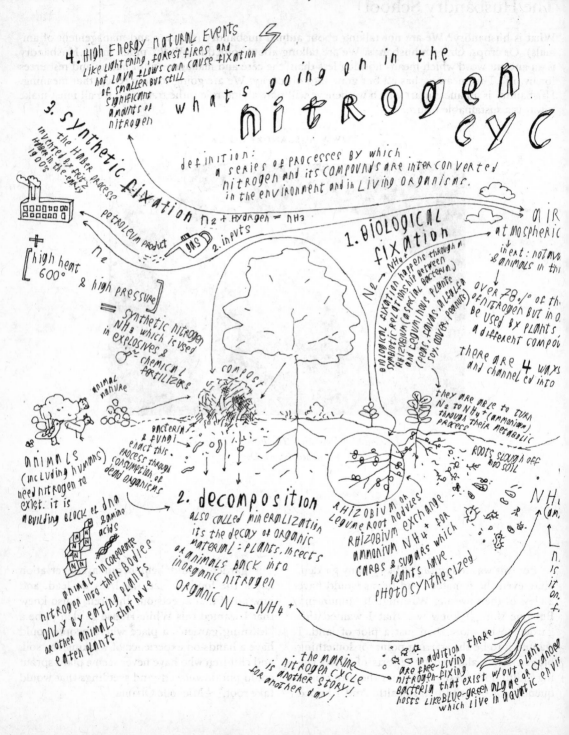

THE FIRST IN A SERIES OF 12 SUCH POSTERS, COLLABORATION BETWEEN BROOKE BUDNER
AND GREENHORNS, FOR SCHOOLS AND EDUCATORS.
LITTLECITYGARDENS.COM

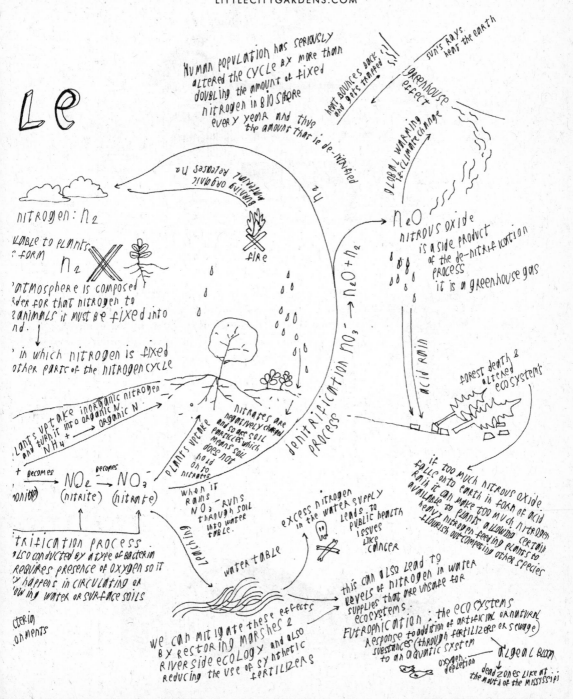

FARM BUILDINGS.

HRIFTY farmers keep two leading points in view—first, to raise all they can from their lands ; and secondly, to take good care of these products after they are raised. Or, as the old maxim has it in reversed order, they "keep all they can get, and get all they can;" not applying it, however, to their intercourse with men, but with their farms only.

They get all they can, by preserving and increasing the permanent fertility of their fields ; and they keep all they get, by not wasting their crops from a want of shelter, nor their flocks by exposure to storms and cold.

The Illustrated Annual Register of Rural Affairs
and Cultivator Almanac for the Year 1862,
Thomas, J.J. (John Jacob) www.archive.org

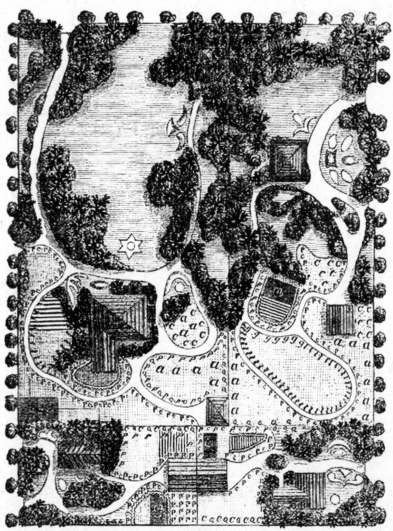

Fig. 2.

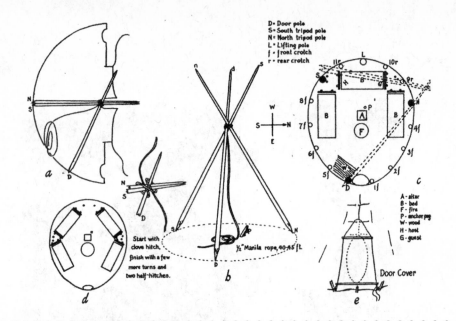

D = Door pole
S = South tripod pole
N = North tripod pole
L = Lifting pole
f = front crotch
r = rear crotch

Start with clove hitch, finish with a few more turns and two half-hitches.

½" Manila rope, 40-45 ft.

A - altar
B - bed
F - fire
P - anchor peg
W - wood
H - host
G - guest

Door Cover

a   b   c   d   e

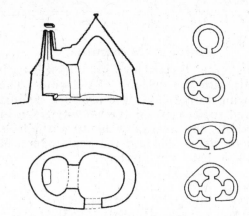

A transitional form of *trullo* field shelter, nonrectilinear in plan.
A main space is joined to a smaller one housing a fireplace. Such
shelters are often found with two or three smaller spaces attached
to the main space, as illustrated in the small sketches. (Plan and
section courtesy *Byggekunst*, redrawn by the author.)

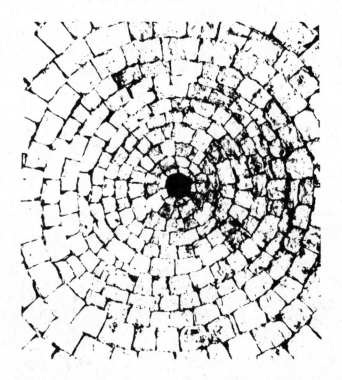

Force of Windmills.

The force exerted by windmills will vary greatly with the velocity of the wind. The following table shows the pressure against a fixed surface; from the velocity given in this table, the average velocity of the sails must be deducted, and the remainder will show the real force exerted:

Miles an hour.	Pressure in lbs. on square ft.	Description.
1	.005	Hardly perceptible.
2	.020 }	Just perceptible.
3	.045 }	
4	.080 }	Light breeze.
5	.125 }	
6	.180 }	Gentle, pleasant wind.
7	.320 }	
10	.500 }	Pleasant, brisk wind.
15	1.125 }	
20	2.000 }	Very brisk.
25	3.125 }	
30	4.500 }	Strong, high wind.
35	6.125 }	
40	8.000 }	Very high.
45	10.125 }	
50	12.500	Storm or tempest.
60	18.000	Great storm.
80	32.000	Hurricane.
100	50.000	Tornado, tearing up trees and sweeping off buildings.

The Illustrated Annual Register of Rural Affairs
and Cultivator Almanac for the Year 1862,
Thomas, J.J. (John Jacob)
www.archive.org

"A happy sign in this troubled year is the conviction held by many people that the time is now and the chance is here to plan for a more secure agriculture in the United States."...They apply different words to the goal they mean to reach. Some call it a balanced agriculture or permanence in farming. Some call it stability or continued abundance. Others think of it as converting agriculture from a wartime to a peacetime basis. But whatever the words, they express a common idea- the goal of security. To the farmer, security means year to year and generation to generation assurance that he can use his land as it should be used, free from fear of boom or bust, that he will have a fair market for the products of his soil and toil, that he will get the amenities that he earns, that he can serve community and country. And any of the people with whom I have talked look upon grassland as the foundation of security in agriculture. They believe in grass, and so do I, in the way we believe in the practice of conservation, or in good farming, or prosperity, or cooperation. For grass is all those things; it is not just a crop. Grassland agriculture is a good way to farm and to live, the best way I know of to use and improve soil, the very thing on which our life and civilization rest."

– Alfred Stefferud

FROM THE YEARBOOK OF USDA ON GRASS

To be fully effective, grassland agriculture, like agriculture, is subject to modification by the sweeping influence of social and economic change. It cannot deny or ignore the effects of varying systems of land tenure, the requirements of sound credit, the impact of shifts in national policies, or the political trends resulting from these and other factors that affect the lives of people on the land.

What the soil may produce, or may be made to produce, this year or next is of immediate importance in terms of current food supply and demand. How the soil may be used to insure sustained production is of continuing importance to national welfare. How to use his soil for his immediate purposes, and yet keep it usable indefinitely, is the users' responsibility and trust.

Dogmas grow like potatoes - creeds and carrots, catechisms and cabbages, tenets and turnips.

The direct agency upon which all these conditions depend, and through which these forces operate, is food. Temperature, humidity, soil, sunlight, electricity and vital force express themselves primarily in vegetables' existence. They furnish the basis of animal life that yields sustenance to the human race. What a man, a community or nation can do, think, suffer, imagine or achieve depends upon what it eats.

The primary form of food is grass. Grass feeds the ox, the ox furnishes the man, and man dies and goes to grass again. And so the tide of life, with everlasting repetition in continuous circles, moves endlessly on and upward. In more sense than one, all flesh is grass.

- John James

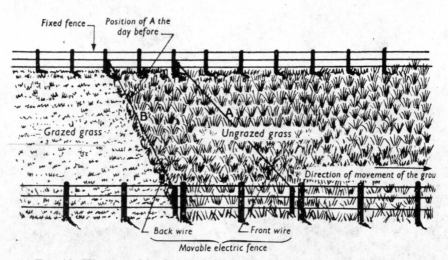

FIG. 12. The area of fresh grass is limited by two electric wires (front and back) which are moved along two fixed fences.

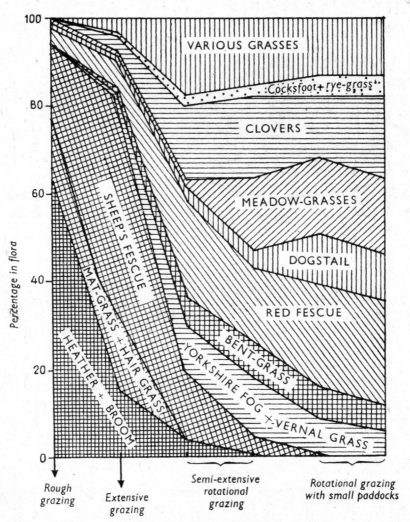

FIG. 34. Influence of method of grazing on the improvement of deteriorated flora.

From Klapp (70), Fig. 74, p. 221.

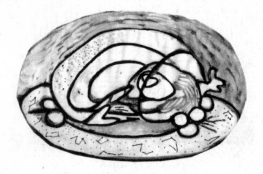

If part of that constellation
We might, orange hued,
Herd the blazing soot
Out of the Eastern sky

–Mariette Lamson

NOVEMBER

———

DOWNPOUR

NOVEMBER 2013

zodiac signs

Cancer — Good for planting, irrigating, grafting – transplanting / Most fruitful
Scorpio — Unequaled for vine-type growth / Fruitful
Pisces — Good for root crops but a very poor time to make manelines / Fruitful

Taurus — A time to plant lettuce, cabbage, sturdy stalks – root crops / Semi-fruitful
Capricorn — Good time to plant silver – root crops / Semi-fruitful
Libra — Time for working of hay, grain, flowers and crops where beauty is concerned / Semi-fruitful

Aries — Time for cultivating / planting where weed destruction is sought after. Cleanses botched and is rule of the poison / Semi-barren
Sagittarius — Semi-barren
Aquarius — Suitable for weeding / destroying of unwanted growth / Barren

Virgo — Good for bringing color to the garden / general cleanup
Gemini — Good for cultivating / destroying unwanted growth
Leo — A time for killing weeds – unwanted growth. Color should be drawn off for stumps. A good time to cut hair – if curly hair is desired, cut hair in the increase of the moon.

8 — ♐ ♑ F — ☀ 6:35 16:44

1 — Harvesting of root crops should be avoided on leaf days, as they are prone to rot. Today is a root day, and thus an optimal time to harvest root crops for storage. — ♍ F ☀ 7:27 17:52

9 — ♑ SA ☀ 6:36 16:43

16 — ♈ SA ☀ 6:44 16:37

17 — The Leonid Meter Shower peaks today. As there is a full moon, it will harder to see, but some meteors may be visible after midnight. — FULL ♈ ♉ SU ☀ 6:45 16:36

24 — ♐ ♌ SU ☀ 6:53 16:32 ♏

25 — AG ♌ M ♏ ☀ 6:55 16:31

2 — ♍ SA ☀ 7:28 17:51

10 — PG ♑ ♒ SU ☀ 6:37 16:42

18 — ♉ M ☀ 6:47 16:36

26 — ♌ TU ♏ ☀ 6:56 16:31

3 — NEW ♍ ☰ SU ☀ 6:29 16:50

11 — 1887, Italy: Cittadella Colony, a co-operative agricultural association, is founded by anarchist Giovanni Rossi. — ♒ M ☀ 6:38 16:41

19 — ♉ TU ☀ 6:48 16:35

27 — ♌ ♍ W ♏ ☀ 6:57 16:31

4 — ♍ M ☀ 6:30 16:49

12 — ♒ ♓ TU ☀ 6:40 16:40

20 — Signs of snow: +When fleas are in abundance +When the grouse drum at night +When the Thanksgiving turkey has thick feathers +When the hen molts before the cock +When pigs squeal in winter (blizzard) — ♉ ♊ W ☀ 6:49 16:34

28 — ♍ TH ♏ ☀ 6:58 16:30

5 — In 1913, Los Angeles receives its first piped-in water from 200 miles northeast of the city. J.B. Lippincott convinced local farmers to relinquish their water rights to him, casting an impression that he would use the water to improve the valley. — ♏ TU ☀ 6:31 16:47

13 — ♓ W ☀ 6:41 16:40

21 — ♊ TH ☀ 6:50 16:34

29 — If on the trees the leaves still hold, The coming winter will be cold. — ♍ F ♏ ☀ 6:59 16:30

6 — ♏ ♐ W ☀ 6:33 16:46

14 — ♓ TH ☀ 6:42 16:39

22 — Castrate and dehorn animals when the Moon is waning for less bleeding. — ♊ ♋ F ☀ 6:51 16:33

30 — ♍ ♏ SA ☀ 7:00 16:30

7 — 1932: "Stay at Home–Buy Nothing–Sell Nothing" Midwestern farmers affiliated with The Farmers' Holiday Association endorsed the withholding of farm products from the market, in essence creating a farmers' strike. The group urged farmers to declare a "holiday" from farming. — ♐ TH ☀ 6:34 16:45

15 — ♓ ♈ F ☀ 6:43 16:38

23 — ♋ SA ♏ ☀ 6:52 16:32

I Worried About The Gap

by ROSMARIE WALDROP

I worried about the gap between expression and intent,
afraid the world might see a fluorescent advertisement
where I meant to show a face. Sincerity is no help once we
admit to the lies we tell on nocturnal occasions, even
in the solitude of our own heart, wishcraft slanting the
naked figure from need to seduce to fear of possession.
Far better to cultivate the gap itself with its high grass for
privacy and reference gone astray. Never mind that it
is not philosophy, but raw electrons jumping from orbit to
orbit to ready the pit for the orchestra, scrap meanings
amplifying the succession of green perspectives, moist
features, spasms on the lips.

"Butternut Squash"

GW

He who receives an idea from me, receives instruction himself without lessening mine; as he who lights his taper [(candle)] at mine, receives light without darkening me.

That ideas should freely spread from one to another over the globe, for the moral and mutual instruction of man, and improvement of his condition, seems to have been peculiarly and benevolently designed by nature, when she made them, like fire, expansible over all space, without lessening their density in any point, and like the air in which we breathe, move, and have our physical being, incapable of confinement or exclusive appropriation.

-Thomas Jefferson

Thanksgiving

by SARAJANE SNYDER

Reviving the lost art of daily thanksgiving is yet another job for all of us young farmers and agrarians who have taken on the radical task of revitalizing our culture. Like so much of what we're setting out to revive and renew, the simple act of giving thanks for what we've got is a means of nourishing our communities in a way that was once second nature to humans living intimately with the land.

In just a moment of contemplation it can become obvious to anyone working with living systems that it is a complexity of birth, death, and transformation that allows for our existence. Soil, seeds, rainfall, decomposition, springtime, lambs, apples—I dare you to study any of these carefully, with all of your senses engaged, and not arrive on the doorstep of wonder and appreciation. I'd be willing to wager that even the most learned biologist could not sincerely contemplate a 200-year-old oak for 10 minutes without arriving at a space, however tiny, of not-knowing—of mystery. In that little fingernail sliver of not-knowing, not-explaining, not-understanding is where we can find the vast sacred place where all things belong. Something tells me this is the place where gratitude is rooted.

To revive the tradition of blessing and thanksgiving, however, one need not be a holy agent of the mysterious or divine (any more than you already are, at least). And one need not even believe in or worship a creator in order to appreciate being part of creation. One only needs to look around with a moment of intention and a quiet mind before gratitude naturally arises.

Thanksgiving serves to create space where we can connect with each other and to the world we live in. Giving thanks, either privately or with others, imbues our daily activities with love and it roots them in the world of the living. There are many reasons to give thanks at the beginning and close of the day's work, or at mealtimes, or even moments of great importance in the farm season: turning in the first covercrop of the spring, the birth and death of animals, sowing the first flats in the greenhouse, or the close of the last farmers' market. The Haudenosaunee People practice a form of thanksgiving known as The Words Before All Else. These words preface all tribal ceremonies, meetings, and general events, summoning the gathered assembly to come together as one mind in gratitude for life in all forms and directions. Gratitude is given in turn to earth, water, sun, moon, weather, stars, ancestors, and plants and animals of all kinds.

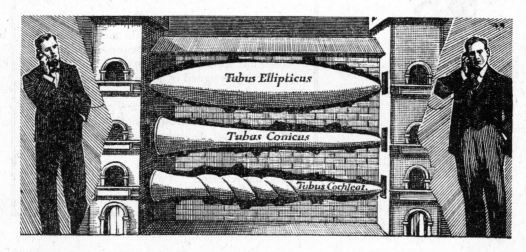

Tubus Ellipticus

Tubus Conicus

Tubus Cochleat.

*Diana, goddess of the hunt, wild
animals and woodlands*

KWR

For some, it can be nerve-wracking to practice thanksgiving with others out loud. If so, try taking a moment of silence where you and each person can summon a particular gratitude of that moment. Eventually you might feel comfortable sharing them out loud. If you find yourself caught by the beauty of the sky or the delicious smell of basil in the sun or the merriment of bees on sedum or the amazing success of your Saturday market, allow that to bloom into what it is: an expression of gratitude. Or, on the other hand, if you find yourself wet in boot & sock with a stuck truck in rainy mud, no one to help you, and your belly rumbling for lunch which was an hour ago, it may be that even there you can find a sweet pip of appreciation by encouraging your senses to overpower your thoughts for a moment.

If nothing else, for those of us who are doing the work we love, we can find reason to give thanks in remembering and recognizing all that had to occur in order to bring us where we are today. As we come together and find others pursuing the dream of positively impacting our food system, let us make thanksgiving practice an essential ingredient in what it is to be a successful farmer.

Sea Vegetable Kraut

by AUDREY BERMAN

Ingredients:
Cabbage (purple variety) -
a large head or 2 small heads
Kosher or Sea Salt - about 3 Tbsp.
A few varieties of seaweed. (I like kombu, dulse, wakame, but you can use whatever types you prefer.)
Sesame Seeds
Ginger
Garlic

**The only thing you really need for kraut is cabbage and salt. Feel free to add or subtract any of the other ingredients. Here's what I've used in other versions: kohlrabi, beets, jalapenos, radish, shiso leaves, fennel seeds, caraway seeds, elderberries, juniper berries, black peppercorns, bay leaves, poppy seeds...

Tools:
your hands
a big mixing bowl - stainless, glass or ceramic - no plastic please (to prevent leaching)
a crock or large ball jar - to pack the kraut into for fermentation

Steps:
Start by chopping up cabbage. As you chop, throw into your large bowl and start adding salt as well. Squeeze / scrunch the cabbage with your hands in the bowl. Continue doing this until all cabbage is chopped and scrunched in the bowl. You can also pack it down with your fist. The goal is to release the juices from the cabbage with salt and pressure. Use no more than 3 Tbsp. of salt for a large head of cabbage. If it's small use less, if you do two large heads use a little more. But don't over salt it! I say a little goes a long way, especially with the natural salts found in the seaweed.

Chop up the other vegetables you want in the kraut and add them to the bowl, mixing and squeezing everything together.
Add seaweed and seeds, mix in.

Now you want to take that mixture and pack it into a crock or jar. You really have to pack the cabbage in tightly so that it's very dense in the container. This allows the water to pull out of the mixture and submerge the kraut.

Once your vessel is packed, if it's a ball jar, I like to take a smaller jar or mug, fill with water and place within the top of the ball jar so as to add weight to the kraut. You can do this with a plate and a rock, or any number of things. The goal is to have the kraut submerged until its own juices so as to prevent molding, etc. Place a dishtowel over the whole thing to keep out any unwanted creatures.
After the first 24 hours you will notice that the juices have leeched out and covered the sauerkraut, either partially or all the way.

From now on, for about 5 days to a week, you need to check the kraut every day, skim off any mold that has formed on the surface (this typically happens when the water is not fully covering the cabbage, but it's totally normal, not to worry) and press the mixture down with smaller jar or mug or your fist. This is important because it squeezes out more juices and helps to keep the whole mixture under water so it can evenly ferment.

After 5 days, taste it. If it's not strong enough or still too salty, leave it out for a day or two more and taste it again.

Refrigerate after it's done to slow down the fermentation process, but it will get zestier over time.

And ta-da, SAUERKRAUT!

JER

RENNET

Is unquestionably one of the most important things in cheese-making. We can separate the caseine from the whey by the use of acids, and even by the natural process of souring, but the product is not that mellow, rich and palatable substance known as cheese. So it is said by the chemists, that the spores, or seeds, of the blue mold are identical with the active properties of good rennet; but no one has yet succeeded in making a marketable article by the use of blue mold instead of rennet. Practically, whatever theory may show, we have no substitute for the active properties of the stomach of the calf, in cheese-making. The soakings of the stomachs of the young of other animals—as the pig, the lamb, the kid, etc.—will cause coagulation, and the extract from the stomach of the pig is said to be stronger than that from the stomach of the calf; but these have never been used to any considerable extent, and we have no knowledge of experiments to satisfactorily determine the relative value of the stomachs of the young of various species of animals for the purposes of the cheese-maker.

At present, therefore, we must confine ourselves to the saving and preparation of the stomach of the calf. Of course the true or digestive stomach—sometimes called the "second stomach"—is the one to be saved. This should be healthy and active, and ought to be saved at the stage when it is just fairly emptied, and the secretions are copious, causing a keen appetite on the part of the animal. The calf should not be less than three days old, and probably ought to be five or six days old, so that all the organs may be in active and vigorous operation. It should go without eating, immediately before being killed, for twelve to eighteen hours. A good way is to feed at night, muzzle the calf or put it where it cannot lick dirt or get hold of straw, hay or other solid substance, and kill it some time during the next forenoon. The stomach should be removed from the calf as soon after killing as possible, as decomposition begins very soon, and goes on very rapidly among the warm vital organs when life has departed. The stomach should be turned inside out and carefully cleared of all

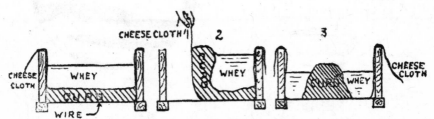

FIG. 553. How to hasten draining of cheese curd. Lift the cloth first at one side of the draining board and then at the other, so as to pile the curd in the centre and let the whey run off more freely

foreign substances, but not washed, and then well salted and stretched on a bow in a cool, dry place; or salt the ends well, tie up one end, blow up the rennet like a bladder, close the other end, and then hang up to dry. When dry, tie your rennets up in paper bags—flour sacks are as good as anything—and keep in a cool, dry place until wanted for use. Freezing does not injure them, and they are best when not less than one year old.

We are confident that the quality of American cheese would be greatly improved by the exercise of more care in saving and using rennets—never using any that are under one year old, and not perfectly healthy and sweet.

The best mode of preparing rennet is that practiced by Dr. L. L. Wight of Whitestown, Oneida Co., N. Y., who has taken the first premium at the State Fair for two years in succession, for the best five factory-made cheeses. He says: " I take pure sweet whey and steam it to boiling, and remove the scum which rises; then let it stand until it settles, and decant the clear whey, leaving the sediment at the bottom of the cask. When this whey is cold and acid, I soak the rennets in it, adding a little salt, but only just sufficient to preserve them from tainting. Of this liquid I use enough to commence coagulating the curd in fifteen minutes.' It requires about one good, strong rennet to each gallon of whey thus prepared. The mode of scalding the whey, so it is not scorched, is of no consequence.

LH

EDITORIAL NOTE: Louella has made these as a template for home-cheesemakers. Feel free to appropriate or design your own to use. Hobbies can become career changes, so have as much fun as possible while you learn.

The Sweetness of Soured Milk: Reasons To Craft Your Own Cheese

by LOUELLA HILL

Humans are an opportunistic species; we've been manipulating fellow mammals for thousands of years, thirsty for their milk, hungry for their meat, frozen without their hides and weak without their strength. We've led herds of goats and sheep over mountains and through valleys. While the animals fed on grass, we gulped their milk. When there was extra, we turned it into dried milk nuggets.

Many centuries later we remain ravenous for milk from our generous friends. And luckily we've moved beyond dried nuggets. We've mastered the art of stabilizing milk, both for storage and flavor purposes, converting it into anything from Comté to caramel. Perhaps we love milk so much for this reason - it offers us flavors that near nirvana. Maybe we're innately tuned into milk because of its nutrient density - what opportunistic species wouldn't covet such a food source? Or maybe we love it because it was our first taste of life.

Whatever the case, I'd like to present the idea of culturing dairy at home. Learning how to convert milk into yogurt, kefir, butter or cheese is an act that reaches simultaneously towards the past and the future. Being part of where your food comes from means a more nourished, more localized, more ecological way of living. Plus, homemade cheese projects can taste wonderful. I've been making cheeses professionally and in my home kitchen for a decade now, all the while ruminating on why making cultured dairy is so cool:

SUPPORT HAPPIER HERDS   When you grab a block of cheese from the supermarket, there's no connection to the animals. As a home cheese maker, however, the quest for really fresh, high quality milk might drive you to make new friends. If you're lucky enough to find a nearby farmer willing to sell, you'll also (hopefully) be paying a more honest price. Milk is the country's most underpriced beverage, with the price paid at the supermarket a far cry from the true cost of its production. After buying directly from the farmer and adding your own labor to the mix, you'll soon understand why cheese could be mistaken for gold and why good milk is really precious. By increasing demand for more localized, more balanced dairying, we start to move away from our modern industrial dairying monstrosity.

DISCOVER NEW FLAVORS   We've realized apples are more complex then red, yellow and green. But what about cheese? Can't we go beyond blocks of cheddar, mozz and swiss? The majority of cheeses we eat are mass produced, their flavors muted to have widespread, generic appeal. By making your own, you open the door to a universe of taste. Maybe your cows are Ayshires instead of Holsteins. Maybe your goats frolicked in the garlic patch. Are the walls of your root cellar rich with mold spores? What happens if you put local mead on an aging goat cheese? By putting together a unique set of circumstances - animal, health, diet, environment and so on - you have an opportunity to experience flavors new not just to your mouth but possibly to humankind as well.

REDUCED PLASTIC EXPOSURE We know not to microwave plastic cups or use bottles with BPA. But did you know many dairy products, including sour cream, stirred yogurt, cottage cheese, cream cheese and ricotta are 'hot packed'? That means we pour piping hot, slightly acidic cheese slurries into low-grade plastic containers in order to extend shelf life. This offers advantages to the producer but what does it mean for you?

GO ZERO-WASTE Do you buy bulk everything, bring your own bags to the farmers' market, even take time to refill containers with olive oil, shampoo and soy sauce at the coop? If so, you probably also have a large stack of 32 oz. yogurt containers hanging out in some cupboard. Yogurt is rarely sold in bulk and rarely sold in returnable glass containers. The solution? Make your own! Purchase milk directly from a farmer or in a returnable glass bottle. Turn the milk into yogurt and have nothing left to throw away!

EXTEND SHELF LIFE You're headed out of town for a week and know that the gallon of milk you just bought won't be good by the time you get back. You could pour it down the drain or pawn it off on a neighbor. Or you could turn it into luscious buttermilk and gain nearly a month of shelf life. All you need to have on hand is freeze-dried culture. The idea of soured milk will now be rather sweet.

SAVE SOME MONEY Though it won't always be cheaper to make your own cheeses (there's just a 8-9% yield on aged cow milk cheeses), it will be cost effective to make your own yogurt, kefir and buttermilk (which has 100% yield). If a quart of good yogurt costs $6 and a gallon of organic milk costs $6, by turning the milk into yogurt yourself you'll get four tubs of yogurt for

the price of one, assuming you use cultures from previous batches.

ADD NOTHING EXTRA When you buy mass-produced dairy products, you're likely purchasing a few things extra. Processing aids and stabilizers include anything from guar gum to pectin to formaldehyde to titanium dioxide. Sometimes these are not even being mentioned on the label. Extra ingredients help produce a prettier, more stable product. But for those of us wanting to eat simply, the only way to ingest just milk and cultures may be to make your own.

LOCALIZE YOUR ECONOMY Enjoying milk from cows on the edge of town offers many benefits. You're more likely to protect a farm if you can see it. You're more likely to eat foods at peak ripeness if they aren't picked to travel. Even better, by purchasing locally, you're strengthening the community and the economy in which you live in. Once you've gotten the hang of home cheese making, you can take 'local economy' to the next level with bartering. Fill your fridge with fresh yogurt and barter away! I've turned dairy products into haircuts, photographs, vegetables, knives, consulting work, babysitting and plumbing repairs.

SM

GROW A GREAT GUT The human gut microbiome, which has evolved with us over many millennia, may be better served when we eat fermented or living foods. Long before we were chasing goats up and down mountainsides, we were bacteria-rich dinner plates. The sterilized and stabilized foods of the modern era are recent occurrences. Studies suggest the benefits from probiotic foods, such as yogurt and kefir, may include the suppression of cancer, insomnia, and diabetes. The probitic potential of yogurt and kefir is most pronounced immediately after it has cultured (not in week 12 when you might buy it from the supermarket). Some people believe probiotics may even lead to longer life. At the Reading Terminal Market in Philadelphia I saw a sign that read, "Drink buttermilk, live forever." I'll edit it only by adding, "Make your own buttermilk, drink it, and live forever."

Louella Hill, a.k.a. The San Francisco Milk Maid, teaches home cheese making courses in the Bay Area. She's currently working on a cheese making manual called "Kitchen Creamery". It's due out Fall 2014 by Chronicle Books.

RESCUE

close to a wall whose bricks were all
Rain-soaked and old,
A little cat hunched where he sat
Dripping and cold

deaf to his cry, crowds hurried by,
Past his me-ow.
I couldn't bear to see him there:
he's my cat now.

We have long talks; Lincoln-Park walks
Cream puffs and pie
A windowpane shuts out the rain
Keeps my cat dry.

Days would be dim
So says my cat-
Had I let him
Sit where he sat.

B.D.F.

RECIPES

by GARTH BROWN

ROAST BEEF

Ingredients:
1 roast from a fat cow, preferably from the loin or rib & Salt

Let us assume it is the sort of day in which the maple's leaves, wild and red in their death, pirouette singly and in threes from the fingers that held them tight since their unfurling. The roast sits on the butcher's block in porcelain dish, and the blood from the meat has darkened the hairline crack across the bottom. Soon it will leak, but for today, for this, it is sound enough. A gust rattles the window in its frame, and for an instant a single leaf presses against the glass, backlit aflame by the sun, and then it falls.

After you awoke, before you made the coffee, you removed the roast from the refrigerator, sprinkled salt across its top, and rubbed its entirety, which brought unbidden memories of rubbing sand against your thigh when you were seven and visited Cape May for the first time. You wanted to leave after an hour, but your parents made you stay for the entire afternoon, even though they spent the whole time fighting. The gulls circled, and their calls were mocking in your ears. Out of spite and boredom you took a handful of wet sand and placed it on your leg and deliberately massaged it until the skin was red and raw.

· ·

CUTS SUGGESTION LIST
by Adam Danforth

COMMON CUTS	AFFORDABLE OPTIONS	ALSO KNOWN AS	PREPARATION
FLAP MEAT	FLANK		BARBECUE
	SIRLOIN FLAP	BAVETTE	BRAISE
	SKIRT STEAK	INSIDE/OUTSIDE	BROIL
	TOP ROUND CAP	SKIRT	GRILL
			PAN FRY
GENERAL	BOTTOM ROUND		PAN ROAST
CHEAP CUTS	HEART	OUTSIDE ROUND,	ROAST
	HEEL	FLAT	
	NECK		
	OXTAIL		
	SHANK		
	TONGUE		

Last night your neighbor Percival - Percival! - returned in his red Camry followed by three other cars, their headlights scouring the darkness from the gravel drive you share, their raucous calls rising through your open bedroom window, like the gulls all those years before, though, to be honest, you didn't think of that at the time, and they didn't actually sound that much like gulls. It would be a nice symmetry if they had.

By lunch Percy and his cadre have begun grilling, and the smell of lighting charcoal now perfumes the dining room. You have made another pot of coffee, which is more than you usually drink, and between jittery sips you outline three possibilities.

1.) You could turn the oven on as high as it goes and blast the roast for a few minutes before dropping the temperature. But this never provides a satisfactory crust. The variation, so much in vogue these days, promulgated, you have no doubt by Cook's Illustrated and the increasing popularity of sous vide cooking, would be to cook the meat until nearly finished, and then to raise the temperature. This is too cute by half, and in your experience it often leads to an overdone roast that still does not have an ideal crust

2.) You could eschew convention and simply roast the meat at a low temperature, accepting that some complexity of flavor will be lost. In return the meat would have an unparalleled evenness of color, a most toothsome texture, indeed, a sublime unity. It would also make the best leftovers tomorrow.

3.) You stand up from the table and finish the coffee. A circle of grounds outlines the slight declivity where the mug's base meets its side. You take the roast in its tray and carry it through the listing screen door. From within Percy's house you can hear screams and laughter, and the cars

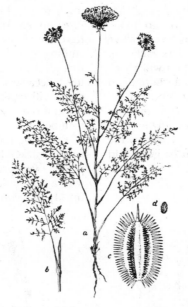

FIG. 81.— Wild carrot; a, plant in bloom; b, leaf; c, seed, magnified; d, seed natural size.

from last night are parked at random across his yard, as if they were the playthings of a toddler who had lost interest and left them where they lay. Percy is alone on the front porch, tending an impressive bed of coals, over which the breasts of three-dozen chickens are steadily charring away to nothing.

"Hey Percy," you say.

"Hey."

"Do you think I could throw this roast on the grill for a second when you're done?"

He looks at you, bleary, hung over, the rings around his eyes purple as thunderheads.

"Uh. No. Sorry, I think maybe some of the other guys want to grill something." He scratches himself and looks at your bare feet and worn jeans and your tan, goosebumped arms, thinking and rethinking, and you meet his unwelcome gaze.

"Actually, it could probably work," he says. "Come back in twenty minutes or so. I'll go ask around, see if anyone else brought anything."

As soon as he is through the door you move the blackened pieces to the grill's periphery, and in three minutes you have seared and flipped and seared the roast. The coals are far too hot for chicken, but you dutifully return the pieces to the center. You would feel guilt for ignoring him, but let's face it, Percy's kind of an asshole.

It doesn't matter that the porcelain dish still has a film of blood across it, for you will go home, and in another hour you will set the oven for 225 and put the roast inside it on a wire rack, not bothering to let it preheat first. You will check the temperature throughout the afternoon, and when it reaches 125 you will remove it to the stovetop, drape a piece of tinfoil over the top, and ball your fists tight as you resist any temptation to cook the meat longer, for you know the centerpiece of your repast should be good and bloody.

"Actually, it could probably work," he says.

"Come back in twenty minutes or so. I'll go ask around, see if anyone else brought anything."

As soon as he is through the door you move the blackened pieces to the grill's periphery, and in three minutes you have seared and flipped and seared the roast. The coals are far too hot for chicken, but you dutifully return the pieces to the center. You would feel guilt for ignoring him, but let's face it, Percy's kind of an asshole.

It doesn't matter that the porcelain dish still has a film of blood across it, for you will go home, and in another hour you will set the oven for 225 and put the roast inside it on a wire rack, not bothering to let it preheat first. You will check the temperature throughout the afternoon, and when it reaches 125 you will remove it to the stovetop, drape a piece of tinfoil over the top, and ball your fists tight as you resist any temptation to cook the meat longer, for you know the centerpiece of your repast should be good and bloody.

STUFFING

INGREDIENTS:

Items too sundry and numerous to list in a comprehensive manner, potentially including: chicken broth, bread, rice, chestnuts, hazelnuts, sausage, sage, mustard, onions, carrots, etc.

The smells of the cooking roast and the warmth radiating from the oven door, in this, the house where your grandmother lived, is a hoary trope, no doubt about it. But, you think, what is cooking other than striving to manifest one stereotype or another? Which is better - making stuffing with, say, blueberries and feta, presumably a rare enough combination that most people will have encountered nothing like it, or to make stuffing of such quality that it nears the Platonic, that all future stuffings will fail to live up to it, until, decades in the future, those still living who ate it will try in vain to convince themselves it could not possibly have been as good as they remember?

From the refrigerator you take the two Ball jars of chicken broth you made from Thursday's picked over carcass, each capped beneath its lid with a yellow skin of fat. They have solidified so thickly that you resort to a spoon to scoop them

NC

into the saucepan, and when you add the wild rice it sits atop until the gelatinous mound melts.

And then there's everything to sauté - an onion, of course, but also a pound of sausage and three stalks of emerald celery, and the sage from the pot outside. Leaden clouds have covered the sun by the time you finish dicing a carrot, and the house shudders in the rising wind so loud you barely notice that the phone is ringing in the living room.

"Hello?" you say, and when you hear your mother's voice you sit down on the edge of the coffee table. She gets into her questions, so predictable you've run down the answers in your head before she's started.

"I have everything taken care of except the wine, I don't have a date, yes, Samantha promised she would be on time, I haven't warmed up to the neighbors, yes, it will be nice to have dinner together." You wait for her to ask each in turn, and through the window you watch two of Percy's friends attempt to pass a frisbee. Every throw travels six straight feet before being grabbed from its trajectory by the wind and dashed into the lawn. One of them picks it up and hurls it wildly, and it comes crashing into the window through which you are watching. The glass doesn't break, but you feel a surge of anger as the plastic disk is retrieved with an affected nonchalance. Percy stands on the porch observing, a beer in his hand, and he looks at you and looks away. You beg off of the phone when you smell the onions. They've browned around the edges but haven't burned.

The chestnuts that you gathered and removed from their hedgehog skins and put in a bowl are round and glossy as stones in a brook. Or perhaps they are more like bulbous leather buttons. They also somewhat resemble the polished shells of some terrestrial mollusk. Whatever the descriptive
conceit, they are perfect, yellow-white beneath their mahogany shells when you split them open.

There are two effective ways to separate the edible and inedible portions of a chestnut.

1.) You could cut a cross in the top of each, and then set them on top of a wood stove or in a cast iron pan. After several minutes of regular turning, their shells would begin to peel off of their own accord, and the surface of the nut meat would caramelize slightly. While perfect for eating as a snack, this method has a tendency to dry them out.

2.) After splitting each nut in half, you drop them into a pan of boiling water for a minute. With a slotted spoon you lift them out, and then, with a large pair of pliers, you squeeze each one on its closed end. You work in rounds so that the halves are hot when you get to them, and it is miraculous how easily they slip their skins, miraculous, at least, to anyone who has laboriously scraped the velveteen layer from the wizened surface of a roasted chestnut.

You take the separated pieces and simmer them in chicken broth until they are tender all the way through. Throughout you keep stirring the onions and stirring the sausage and checking on the rice. Once all is ready you consolidate your pans and bowls into one, add two beaten eggs, and spread the mixture in a dish. When the roast is done you will turn the heat higher and bake it, and you will cover the top before the rice dries out. It is not, of course, stuffing in the proper sense, since nothing is stuffed with it. You could call it dressing, as the "Joy" suggests, but that's hardly less confusing. So long as it's memorable, the name matters little.

DESSERT

Ingredients:

Blueberries
Honey
Seasoning (vanilla, cinnamon, salt, etc.)
Slivered Almonds

When the stuffing was out of the oven, keeping warm beneath its tinfoil cap, as your sister carved the roast and your mother set the table, you took a bag of blueberries picked last August and poured them in a buttered dish, spooned honey and cinnamon atop, and finished with a layer of slivered almonds.

And now the dinner is finished, the roast half gone with half remaining for sandwiches, a wedge of brie where a circle was, the stuffing plucked clean of its last handful of chestnuts. You made a salad - fennel, frisee, escarole and the garden's last radishes - and a single purple half lies in the bottom of the cherry wood bowl.

Your mother held forth on her hopes for your romantic life as well as on her ex-husband's (your father's) inadequacies, and her remarks concerning Samantha's eyebrow piercing were extensive enough to nearly cause a premature end to the gathering. But now a strange quiet has settled over the three of you. Sam's fists and jaw have come unclenched, and her eyes are now downcast. Mother is staring at the breakfront, and you imagine she's remembering the innumerable meals, the laughter and the screams that passed across this table in her childhood, though it's more likely she's planning her tomorrow. She has always been one for the future rather than the past. And as for you, you're no longer trying to bring up the shifting of the seasons or whether the oak across the street that was struck by lightening will last the winter. You are here with your mother and your sister, your splintered family, and you have shared a meal for the first time in four years. The room is warm from the woodstove, and the wine is gone, save an inch in each of your glasses.

Into this postprandial reverie the smell of cooking berries insinuates itself, and you tug down the sleeves of your sweater and stand. You have just taken the pan from the oven and set it on a metal Fleur De Lis trivet, a holdover from grandma, when there comes a scuffling on the porch, followed by a tap at the window. You open the door, and Percy is standing there on the other side of the screen.

"Hi," you say.

"Oh, hey Jerry." He clears his throat and scratches at his nose. "I just wanted to apologize for how rowdy things got last night and this afternoon."

"Don't worry about it. I barely noticed."

"Okay. Well, I was also kind of a jerk today, but I was pretty hung over." He shifts from one foot to the other. "I thought I should probably bring something, a gift, but someone got into my wine, so I only had PBR left, and I figured you wouldn't be interested."

"It's fine."

He stands there in the doorway, and you can feel the warmth washing around you and out past him. You step back into the kitchen.

"We were about to have dessert. You want some?"

"I don't know. I should probably get back to my buddies."

"They won't notice you're gone."

You turn around to get the dessert plates, but Samantha has cleared the table and already mother is setting four places. Outside the moon is rising and with it the wind, and the leaves from the maple again swirl against the window in their peremptory flight.

Cream of Celery

Celery is unusually scarce now, and little of it is well blanched and crisp. But even the somewhat inferior quality may be made quite palatable. Cut it into very small pieces, rejecting the toughest green portions. Add only *water enough to keep it from burning, and boil it in a closely covered vessel for an hour, or until perfectly tender. Then add a suf*ficient quantity of milk, first thickened with a tablespoonful of flour to each pint, previously rubbed smooth with two tablespoonfuls of butter, and salt and pepper to the taste, very little of the pepper. Boil and serve as soon as the flour is thoroughly cooked. If made moderately thin with the milk, flour and butter, it can be rubbed through a colander, when it gives a delicious, cream-like soup. Smooth squares of bread well browned are frequently put into the soup when finished. A bowl of this, eaten with bread, the same as bread and milk, makes an excellent noon lunch. *(1882)*

Salad Dressing

It is a great art to make a good dressing for green salad — lettuce or endive. The art consists in mingling the various ingredients so that each will become disguised and in its turn disguise the others while the combination in no wise obscures the delicious flavor of the lettuce or endives — but rather augments it and promotes its digestibility.

Take the yolks of two hard boiled eggs, crumble them with a silver fork or dessert spoon, add about half a teaspoonful of ground mustard and a teaspoonful of salt, and mix all well together. Then add in three portions a dessert spoonful of pure olive, walnut, or poppy oil, and rub the whole to a uniform smoothness. The addition of twice the quantity of oil will improve the salad to the taste of many, and nothing is more healthful; a dash of cayenne pepper, or a few drops of pepper vinegar may also be added; finally add about a dessert spoonful of sharp vinegar, and if the dressing is not fluid enough, a little water or more vinegar, adding it gradually and rubbing thoroughly all the time.

A little experience only is necessary, but it requires tact and patience. Most people abhor oil because they pour it over the green leaves instead of blending it in a dressing. Others douse on oil, catsup, mustard, pepper, salt — every spice or condiment they can lay hands upon — sugar it well besides, and then drown it in vinegar. Think of catsup on a crisp lettuce head — horrible! (1862)

Pleasant and Wholesome Summer Drink

The juice of currants, put up in airtight bottles, affords a foundation for a delicious and wholesome beverage. Put enough water with ripe currants to prevent their burning; heat in a preserve-kettle nearly to boiling; transfer to a bag suitable for straining and press out the juice; add half a pound of clean sugar to each pound of juice and boil, skim, and put up as recommended for putting up the fruit.

The juice alone will keep as well, if not better than the fruit, and, mixed with from one to two parts water, according to the taste, it makes a most refreshing drink. Being entirely free from alcoholic or intoxicating properties, there is no danger of the creation of an appetite for strong drink resulting from its use. It might profitably be kept for sale by druggists at all seasons of the year, and we presume that putting it up for that purpose may be made a source of income worthy of the consideration of currant-raisers who now make wine. Those who usually have more of this abundant fruit than they know what to do with, may find it for their interest to take note of this suggestion. (1862)

Vegetable Soup

Take a good sized chicken, or an equivalent piece of beef or mutton, cut it up and put it in water, rather more than enough to cover it, adding a tablespoonful of salt; boil until nearly tender, skim off the fat; add butter, salt, and pepper, and more water if necessary; then slice into the soup ten large potatoes, one small Swedish turnip, one carrot, two parsnips, (an onion and a few stalks of celery,) with two or three spoonfuls of rice; boil half an hour, or until tender.

Before serving, add a spoonful or two of wheat flour stirred up with cold water. One or two spoonfuls of sweet cream greatly improves the flavor. (1862)

MAKING TINCTURES & SALVES *by* JANA BLANKENSHIP

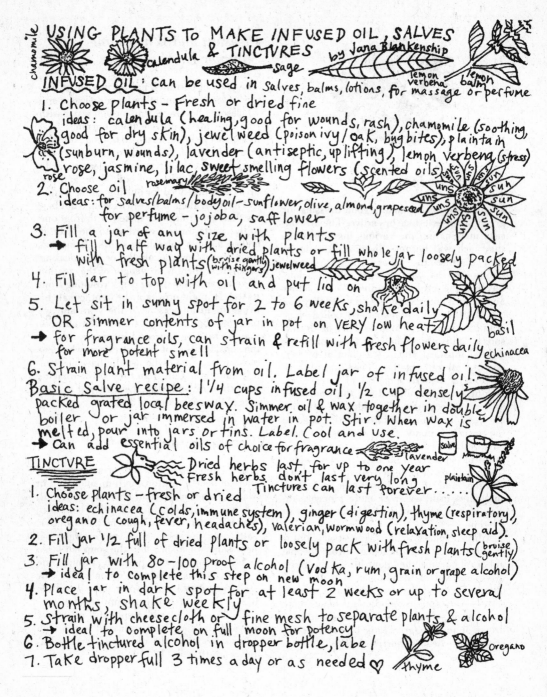

USING PLANTS To MAKE INFUSED OIL, SALVES & TINCTURES by Jana Blankenship

chamomile   calendula   sage   lemon verbena   lemon balm

INFUSED OIL: can be used in salves, balms, lotions, for massage or perfume

1. Choose plants - Fresh or dried fine
   ideas: calendula (healing, good for wounds, rash), chamomile (soothing, good for dry skin), jewel weed (poison ivy/oak, bug bites), plantain (sunburn, wounds), lavender (antiseptic, uplifting), lemon verbena (stress) rose, jasmine, lilac, sweet smelling flowers (scented oils)

rose   rosemary

2. Choose oil
   ideas: for salves/balms/body oil - sunflower, olive, almond, grapeseed
   for perfume - jojoba, safflower

3. Fill a jar of any size with plants
   → fill half way with dried plants or fill whole jar loosely packed with fresh plants (bruise gently with fingers) jewelweed

4. Fill jar to top with oil and put lid on

5. Let sit in sunny spot for 2 to 6 weeks, shake daily
   OR simmer contents of jar in pot on VERY low heat
   → for fragrance oils, can strain & refill with fresh flowers daily for more potent smell

basil   echinacea

6. Strain plant material from oil. Label jar of infused oil.

Basic Salve recipe: 1¼ cups infused oil, ½ cup densely packed grated local beeswax. Simmer oil & wax together in double boiler or jar immersed in water in pot. Stir. When wax is melted, pour into jars or tins. Label. Cool and use.
→ Can add essential oils of choice for fragrance

salve   lavender

TINCTURE   Dried herbs last for up to one year
Fresh herbs don't last very long
Tinctures can last forever.....

plantain

1. Choose plants - fresh or dried
   ideas: echinacea (colds, immune system), ginger (digestion), thyme (respiratory), oregano (cough, fever, headaches), valerian, wormwood (relaxation, sleep aid).

2. Fill jar ½ full of dried plants or loosely pack with fresh plants (bruise gently)

3. Fill jar with 80-100 proof alcohol (vodka, rum, grain or grape alcohol)
   → ideal to complete this step on new moon

4. Place jar in dark spot for at least 2 weeks or up to several months, shake weekly

5. Strain with cheesecloth or fine mesh to separate plants & alcohol
   → ideal to complete on full moon for potency

6. Bottle tinctured alcohol in dropper bottle, label

7. Take dropper full 3 times a day or as needed ♡

thyme   oregano

VALUABLE RECIPES.

Potato glue.—Take a pound of potatoes, peel them, and boil them, pound them while they are hot in three or four pounds of boiling water; then pass them through a hair sieve; afterwards add to them two pounds of good chalk, very finely powdered, previously mixed with four pounds of water, and stir them both together.—The result will be a species of glue or starch, capable of receiving every sort of coloring matter, even of powdered charcoal, of brick, or lampblack, which may be employed as an economical means of painting door-posts, walls, pailings, and other parts of buildings exposed to the action of the air.

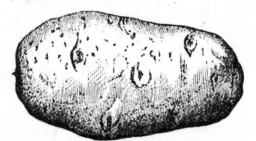

Fig. 10.—*Early Goodrich—Reduced one-half in Diameter.*

Fig. 11.—*Calico—Reduced one-half in Diameter.*

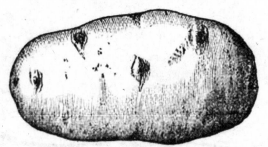

Fig. 12.—*Harison—Reduced one-half in Diameter.*

RECIPES *by* SONYA MONTENEGRO

FINGER-LEEKIN' GOOD DIP

4½ Tablespoons butter
1¼ lbs. leeks, sliced
1 Large potato
1 lb. broccoli (cut stems into pieces)
6 Tablespoons whipping cream
1 Tablespoon fresh dill

Melt butter in a large skillet over medium heat, stir in leeks and cook for about 15 mins - or until soft.

Quarter potato and boil salted water for about 15 mins. Add broccoli stems for 5 mins, then florets for 2 mins more (or drop in florets when potato + stems are tender, cut heat and get food processor ready.)

Drain and mix into food processor. Add cream and dill, season with salt and pepper to taste.

ENJOY.

as a dip with:
fresh veggies
crackers
bread sticks
toast
or....?

EAT SOME PRESERVE SOME FOR NOW LATER

CHUTNEY

8 CUPS APPLES - CHOPPED
6 CUPS ONIONS - CHOPPED
½ CUP GINGER - MINCED
¾ CUP VINEGAR
1½ TBLS WHOLE CUMIN SEED
1½ TBLS MUSTARD SEED
2 CUPS SUGAR
3 DRIED CHILI PEPPERS - DICED

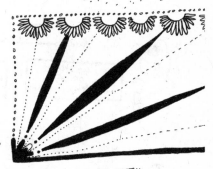

COMBINE ALL INGREDIENTS IN A LARGE POT — COOK + STIR UNTIL APPLES ARE SOFT + ONIONS ARE TRANSLUCENT. TASTE OCCASIONALLY AND ADD MORE SUGAR AND/OR SALT AS NEEDED.

TO KEEP: IF YOU PLAN TO EAT CHUTNEY SOON AFTER MAKING, JUST FILL WARM JARS WITH CHUTNEY AND STORE IN REFRIDGERATOR

~OR~

IF YOU'D LIKE TO PRESERVE YOUR CHUTNEY FILL STERILIZED JARS, SEAL AND HEAT SEAL IN A HOT WATER BATH.

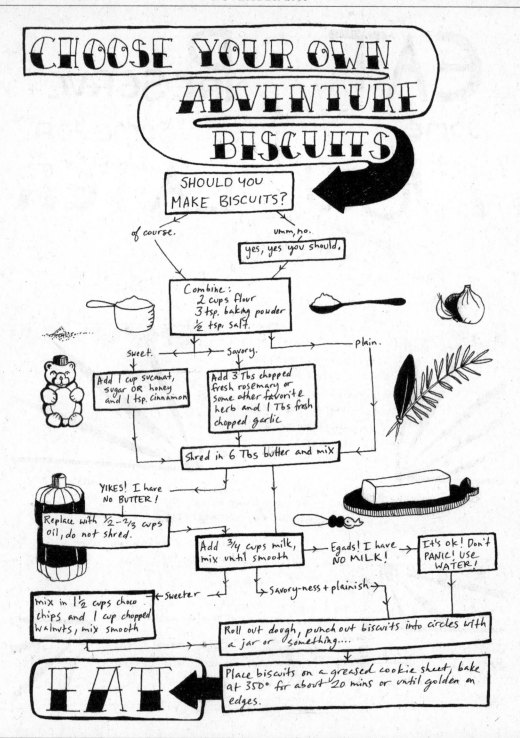

CHOOSE YOUR OWN ADVENTURE BISCUITS

SHOULD YOU MAKE BISCUITS?

of course.

umm, no.
yes, yes you should.

Combine:
2 cups flour
3 tsp. baking powder
½ tsp. salt.

sweet.

savory.

plain.

Add 1 cup sucanat, sugar OR honey and 1 tsp. cinnamon

Add 3 Tbs chopped fresh rosemary or some other favorite herb and 1 Tbs fresh chopped garlic

Shred in 6 Tbs butter and mix

YIKES! I have NO BUTTER!

Replace with ½ - ⅔ cups oil, do not shred.

Add ¾ cups milk, mix until smooth

Egads! I have NO MILK!

It's ok! Don't PANIC! Use WATER!

Sweeter

Savory-ness + plainish

mix in 1½ cups choco chips and 1 cup chopped walnuts, mix smooth

Roll out dough, punch out biscuits into circles with a jar or something....

EAT

Place biscuits on a greased cookie sheet, bake at 350° for about 20 mins or until golden on edges.

TOM'S SPICY GREEN SAUCE

— 3 cups cilantro*
(stems & all)
2 Tbl black pepper
1 Tbl orange peel
2-3 cloves garlic
Fish sauce
and/or
soy sauce

Put all ingredients in a blender and blend. After this initial blending breaks up greens, add fish/soy sauce to make into paste consistency. Add lime juice to taste. Eat on anything & everything!

* Tom advises using any & all greens available, he's used chickweed, arugula, parsley, any mustard greens...

Red fir, white cedar, blue spruce
behind mountain, cold moon
rise of winter solstice hours
of night, worn wool

–Francesca Capone

DECEMBER

GUSTS

DECEMBER 2013

zodiac signs

Cancer — Good for planting, irrigating, grafting – transplanting. Most Fruitful
Scorpio — Exceptional for vine-type growth. Fruitful
Pisces — Good for root crops but a very poor time to make sunshine. Fruitful
Taurus — A time to plant lettuce, cabbage, sturdy stalks – root crops. Semi-fruitful
Capricorn — Good time to plant where next crops. Semi-fruitful
Libra — Time for seeding of hay, grass, flowers and vines where beauty is concerned. Semi-fruitful
Aries — Time for cultivating – plowing where weed destruction is sought after. Chicken hatched one will be carrier. Semi-barren
Sagittarius — Best for destroying unwanted plant life. Semi-barren
Aquarius — Suitable for weeding – destroying of unwanted growth. Barren
Virgo — Good for bringing color to the garden – general cleanup.
Gemini — Great to cultivate – dry – destroying unwanted growth.
Leo — Great for killing weeds – unwanted growth. A time for billing weeds – unwanted growth.

---

1 — SU · 7:01 / 16:29
In 1931 unemployment in America reaches 13.5 million – almost 1/3 of the American work force. 200,000 become freight car migrants on the Missouri Pacific Line.

2 — NEW · ♎ ♏ M · 7:02 / 16:29

3 — ♏ TU · 7:03 / 16:29
1931: Severe drought hits the midwestern & southern plains. As the crops die, the "black blizzards" begin. Dust from the over-plowed & over-grazed land begins to blow.

4 — ♐ W · 7:04 / 16:29

5 — PG · ♐ TH · 7:05 / 16:29

6 — ♐ ♑ F · 7:06 / 16:29

7 — ♑ ♒ SA · 7:07 / 16:29

8 — ♒ SU · 7:08 / 16:29

9 — ♒ ♓ M · 7:08 / 16:29

10 — ♓ TU · 7:09 / 16:29

11 — ♓ W · 7:10 / 16:29

12 — ♓ ♈ TH · 7:11 / 16:29

13 — ♈ F · 7:12 / 16:29
Light from the gibbous moon may make viewing the Geminid Meteor Shower, occurring today through the 15th, difficult. The Geminid radiates up to 60 meteors per hour.

14 — ♈ ♉ SA · 7:12 / 16:29

15 — ♉ Su · 7:13 / 16:30

16 — FULL · ♉ M · 7:14 / 16:30

17 — ♉ ♊ TU · 7:15 / 16:31

18 — ♊ W · 7:15 / 16:31

19 — AG · ♊ TH · 7:16 / 16:31
1871: Corrugated cardboard patented by Al Jones.

20 — ♊ ♋ F · 7:16 / 16:31

21 — ♋ ♌ SA · 7:17 / 16:32

22 — ♌ SU · 7:17 / 16:32

23 — ♌ M · 7:18 / 16:33
In 1975, the Passamaquoddy & Penobscot tribes of Maine win a court decision upholding principle that the US has an obligation to protect the land rights of all tribes, whether recognized by the federal government or not.

24 — ♌ TU · 7:18 / 16:34

25 — ♍ W · 7:18 / 4:34

26 — ♍ TH · 7:19 / 4:25
In 1877 the Workingmen's Party is reorganized as the Socialistic Labor Party. Political activists gain control. It is later named the Socialist Labor Party.

27 — ♍ F · 7:19 / 16:36

28 — ♍ ♎ SA · 7:19 / 4:36

29 — ♎ ♏ SU · 7:20 / 16:37

30 — ♏ M · 7:20 / 16:38

31 — ♏ ♐ TU · 7:20 / 16:39
1890 -- NY: Ellis Island opens, replacing Castle Garden as the American immigration depot.

THE NEW OLD TRADITIONS
THROUGHOUT THE YEAR

YULE: "A Sun is Conceived"
(On/around December 21st)

On or around December 21st we celebrate the festival
or "Sabbath" known as Yule, which marks the Sun's
moment of conception. As the story goes it is on this
day that the Moon and the Night, after a year-long
flirtation finally took the plunge and ignited the Cos-
mos with a naughty sexcapade. So wonderful was this
event, and so much did the Moon want to ravish the
body of Night, that the Moon stretched the Night as
long as she could just to have more to play with. It is in
this way that we have inherited the longest night of the
year, and it is in this way that the Sun was conceived.
Yule is a time when the Moon's guidance of the Earth
into the cozy reflections of Fall and Winter culminate
in a feast.

MAXIM FOR THE YOUNG.—Keep good company or none.

RECIPE - Winter

by ZOE LATTA

make lentils with some bacon grease
serve with jiffy cornbread

Reading (short) List

by CHRISTOPHER CHEMSAK

- *Gaia's Garden* by Toby Hemenway

- *Farms Of Tomorrow* by Trauger M. Groh
  and Steven S.H. Mcfadden

- *The Encyclopedia Of Country Living*
  by Carla Emery

- *Soil Fertility, Renewal And Preservation*
  by Ehrenfried Pfieffer

- *The Unsettling Of America*
  by Wendell Berry

- *Meeting The Expectations Of The Land:
  Essays In Sustainable Agriculture And
  Stewardship* Edited by Wes Jackson

- *Living The Good Life*
  by Helen And Scott Nearing

An Interview With
James Howard Kunsler

by SEVERINE VON TSCHARNER FLEMING

Since the Farm bill failed to pass, we're antici-
pating big cuts in USDA programs. Usually it
plays out that Ag lobbies protect their interests
while conservation programs, food stamps and
less-moneyed interest groups take the big hits.
But if, as you predict, the era of big government
will likely end, ending subsidies for bad farming-
-what would happen then?

JHK– The way we do farming in the USA faces
many destabilizing forces. These include abnor-
mal weather due to climate change, rising price
and scarcity of oil, and the breakdown of distribution networks. Problems with money may
loom largest and most immediate since the biggest "input" for industrial agriculture is capital in
the form of annual revolving loans to big corporate farmers to keep their operations going. We
are entering a period of capital scarcity due to
the impairments of capital formation and debili-
ties of the banking system. The macro trend in all
human economic activities is to downscale and
get more local. We are probably going to have
to grow more food closer to home – not all of it
but at least some. We have no idea how we are
going to get there from the current disposition of
things. The process is likely to be disorderly. The
reassignment of valuable ag land is an enormous
question. And the fact that so much good land
close to towns and cities has been paved over also
presents a conundrum.

What is your forecast for how the young farmers movement might work more in concert
with other groups. (Ie do the Main street movement, smart growthers and transition towns,
land conservationists and doomers.. could they
see big enough to get involved and invest their
money in our farms and agricultural infrastructure?) How do connect strategically with adjoin-
ing movements? What kinds of cooperatives or
innovative economic structures can you imagine

or have you learned about that we should be considering?

JHK - There are themes of both technological
and organizational narcissism that run through
the current public discussion of these issues and
I think this represents the latter - the notion that
we'll organize our way through this set of problems. I think the reality is that the process will be
emergent and self-organizing as "shit happens"
and people in a stressed society have to impro-
vise. Political revolution is one form of impro-
visation, though often a very messy one. I don't
promote it as the best remedy, but it is what happens when societies put off problem-solving for
too long. That said, there is certainly a cohort of
young people who recognize that the future of
farming is not what we've been doing the past
100 years.

Hurricane Sandy, and other weather events,
might well provoke more city-dwellers to consider re-ruralizing. Can you make the case for them
to do so?

JHK - I'm convinced that we will see striking
demographic shifts. The suburbs really have
no future, and fantasies about turning all the
lawns into gardens are naïve. However, our giant metroplex cities will also get into trouble. In
my opinion, they will have to contract substan-
tially. This contradicts the current popular fantasies put forth by Harvard's Ed Glaser and New
Yorker writer David Owen about the glorious
future of the skyscraper city. They are completely wrong about this. Rather, we're going to see
the re-activation and re-population of America's
small towns and small cities – places scaled to the
energy and capital realities of the future – and
places that exist in a meaningful relationship
with food production. The rural landscape will
also have to be inhabited differently, largely because 1) we won't be traversing the landscape in
automobiles and 2) food production will require
much more human attention.

What kinds of human social technologies does
my generation currently lack? How can we re-

learn older ways of " relating"?

JHK – If you are saying how do we recover from the catastrophe of Facebook and iPhones then I believe these are transient phenomena that will disappear when we start having trouble with our decrepit electric grid, which is sure to happen in the years ahead. When these addictive and time-wasting "technologies" fade away, human beings will recover their natural social abilities.

Can you talk about how the end of Cheap energy affects Big Ag, will we be growing sunflower oil for our tractors in vermont? Do you predict the deforestation of the east coast to reclaim agricultural lands, for firewood and lumber?

JHK – These are the kinds of improvisations that belong in the "emergent" folder. We don't know yet. My guess is that we will run far fewer engines than we do now. I worry about deforestation in my corner of the country (eastern upstate New York) though it has already been through one cycle of that in the 19th century. Now we have maturing second and third growth forests where there used to be pastures. We have no idea how people might keep warm a few decades hence. I think there will be fewer people, for one thing.

Cradling wheat

In, The Long Emergency, you predict Protracted Crisis. Farming is the one many of us have chosen to commit to place and community service, but since food is always needed, many of us feel our skills make us more resilient as well.. What other jobs, skills and missions could we consider, for ourselves, for our mates, to be more resilient. What ways of behaving ?

JHK – We face the complete rebuilding of local

networks of economic inter-dependence – i.e. local Main Street economies. This is a massive project that will provide many vocational niches for young people in the future. I'm amazed that so few of you envision the demise of gigantic-scale trade operations of the WalMart variety. This is the way that commerce has been organized for at least two generations, and it seems like a permanent fixture of life, but I assure you the days of WalMart and giant organisms like it are numbered. Their business formulas have no future (e.g. 12,000 mile merchandise supply lines + the warehouse on wheels). As far as behavior goes – I will catch a lot of crap for this – I'm convinced that we will see much more hierarchal social relations in future decades, sharper class divisions, changes in gender relations vis-à-vis the divisions of labor.

It is surprising to many what a large percentage of new farmers are women, women who are interested to work in farm labor, farm management, animal husbandry outside with the boys. What do you make of this?

JHK – Males are psychologically beaten down lately. They were kind of hung out to dry by circumstance. Their choice of vocational roles and meaningful occupations has been limited – especially for the many who are not math whizzes. I think you will see them re-motivated in the years ahead by the recognition that there are many new opportunities in farming and rebuilding local economies. On the other hand, I would not expect an era of the supremacy of women – and there's a lot of fantasy about this these days. Really, the efficacy of women in recent years – e.g. things like greater proportional enrollment in college – can be attributed to the fact that so many jobs of our time (corporate, government) can be done by anyone, male or female. This will not be so much the case going forward. I don't think we have to slide backwards to the days when women were put upon and held back, but their recent victories in the workplace and government have much to do with the demoralization of men.

You've been working hard, writing hard, and speaking cleverly for quite a while trying to wake people up to the stupid decision making and poor planning in our communities. I wish we could call you the Grinch that saved Main street-- but frankly it seems a losing battle. I've recently been reading some of the early works of the Soil Conservation Society, The Land magazine and early writings of Hugh Bennett, Lord Russell, the evangelizers of Soil Conservation. Eventually their programs and policies were adopted by FDR"s Countryside Commission and the New Deal reorganization of the USDA-- you could say they had success. We have a similarly well-heeled, connected movement these days " the local food movement"-- though with less political insiders, but it seems nigh impossible to shift the terms of the debate. Do you recommend street theatre? Is politics perhaps the wrong venue for activism these days?

MUSHROOMS.

JHK - I believe in Schopenhauer's formulation that (I paraphrase) people react to new ideas first by ridicule, then by violently opposing them, and finally by adopting them as self-evident. That's how it will go with our disposition of things on the landscape, the way we organize the human habitat and the farming hinterlands. We're getting closer to that moment when the suburban equation clearly fails. On the urban end of the question, I was inspired to see that new loft apartment buildings (about five stories) are being built in downtown Sioux Falls SD, where I went last month – one of the most sprawl-afflicted cities in the USA. Like all the

Ponzi systems that are running right now, agribusiness, too, will collapse and we'll have to do things differently. I'm very optimistic about the future, but my vision of it is not like many of the reigning popular fantasies.

JHK - If you are saying how do we recover from the catastrophe of Facebook and iPhones then I believe these are transient phenomena that will disappear when we start having trouble with our decrepit electric grid, which is sure to happen in the years ahead. When these addictive and time-wasting "technologies" fade away, human beings will recover their natural social abilities.

Can you talk about how the end of Cheap energy affects Big Ag, will we be growing sunflower oil for our tractors in vermont? Do you predict the deforesation of the east coast to reclaim agricultural lands, for firewood and lumber?

JHK - These are the kinds of improvisations that belong in the "emergent" folder. We don't know yet. My guess is that we will run far fewer engines than we do now. I worry about deforestation in my corner of the country (eastern upstate New York) though it has already been through one cycle of that in the 19th century. Now we have maturing second and third growth forests where there used to be pastures. We have no idea how people might keep warm a few decades hence. I think there will be fewer people, for one thing.

In, The Long Emergency, you predict Protracted Crisis. Farming is the one many of us have chosen to commit to place and community service, but since food is always needed, many of us feel our skills make us more resilient as well.. What other jobs, skills and missions could we consider, for ourselves, for our mates, to be more resilient. What ways of behaving?

JHK – We face the complete rebuilding of local networks of economic inter-dependence – i.e. local Main Street economies. This is a massive project that will provide many vocational niches for young people in the future. I'm amazed that so few of you envision the demise of gigantic-scale trade operations of the WalMart variety. This is the way that commerce has been organized for at least two generations, and it seems like a permanent fixture of life, but I assure you the days of WalMart and giant organisms like it are numbered. Their business formulas have no future (e.g. 12,000 mile merchandise supply lines + the warehouse on wheels). As far as behavior goes – I will catch a lot of crap for this – I'm convinced that we will see much more hierarchal social relations in future decades, sharper class divisions, changes in gender relations vis-à-vis the divisions of labor.

For some time it has been recognized that the soil-erosion problem of the United States could never be satisfactorily solved by Federal action alone. The task is too vast and too complex to be achieved in its entirely by a central governmental agency. While Federal agencies are needed to point the way toward better land use through technical advice and assistance, the initiative and actual work of conservation on a large part of the country's tillable land and rangeland must be undertaken by the farmer and rancher.

It is at once evident, however, that individual efforts to control erosion are likely to be ineffective. They can be costly and can never be anything but piecemeal. The one system of attack on erosion that promises success is the cooperative attack, beginning where erosion begins, at the crests of ridges, and working down, field by field and farm by farm, to the stream banks in the valleys below.

The soil-conservation district, a unit of local government authorized by State law, will speed this type of cooperative action. It is simply a mechanism whereby farmers or ranchers within a watershed, or other natural land use area, may organize for community action and mutual protection in combating the soil-erosion problem. More than half of the States have passed laws permitting local groups of farmers to organize such districts. It is hoped that eventually a significant portion of the National's erodible land will be included in districts.

—Soil Conservation Districts

Trachykele mines in   Trachykele mines
Longitudinal Section   in Cross Section

Louis Bromfield
And His Books

"Bromfield seems to be able to make anything in which he was interested seem like high adventure to those who read of his exploits.

"The book was unquestionably directed toward those who either had a love for the land and for farming or could be influenced to be intrigued by the mystery of the soil. Its appearance was coupled in time with one of the greatest gardening movements in the United States, brought on by the scarcity of food during the war. There were thousands of new small-time farmers in the country, and here was something a thousand times more interesting than a seed catalogue. It was the story of a large-scale attempt by an amateur like themselves to make a bit of land behave properly, an attempt that was carried out with a flourish by a highly imaginative man with plenty of money. However they were told constantly that there was nothing in the process that any ordinary farmer could not do if he had the intelligence and the pioneering courage to try, even without the money" (about Malabar Farm).

LOUIS BROMFIELD BOOK LIST

- *Early Autumn*
- *A Modern Hero*
- *The Rains Came*
- *Mrs. Parkington*
- *The Farm*
- *Malabar Farm*
- *Up Ferguson Way*

Recommended
Children's Books

by PATRICK KILEY

- *A Farmer's Alphabet* by Mary Azarian
- *Hans In Luck* by The Brothers Grimm
- *In My Mother's House* by Ann Nolan Clark
- *Skotny Dvor [Barnyard]* by Evgeny Schvarts
- *Charlotte's Web* by E.B. White
- Beatrix Potter Books

Report Back From The Border

by BROOKE BUDNER, 2010

This summer I took a month off of this urban farming pursuit and went down to the Arizona/Mexico border to work with a humanitarian aid organization called No More Deaths whose mission is to end the deaths and suffering of folks migrating to the United States. The organization is completely initiated and run by dedicated local volunteers and visiting volunteers from around the country. Its work is based on the ideal of Civil Initiative, the belief that communities must organize and take power to uphold humanitarian rights when states or nations cannot or refuse to do so. Every day members of the group hike trails, drop off hundreds of gallons of water and food in remote parts of the desert, and are available on encounter with migrants to provide medical attention. No More Deaths also staffs a desert medical aid tent and Resource Centers in the border towns of Nogales and Agua Prieta. (For more info about the work and mission of NMD check out the July blog post or visit their website: www.nomoredeaths.org)

A complex political and human tragedy is unfolding on our border. It is devastating and confusing to witness even just a slice of it. Every year hundreds of thousands of Mexican and Central Americans set out on a life-threatening journey across mountainous, desert terrain in order to meet family and find work in the U.S. Every year hundreds of these migrants get lost, injured, raped or attacked along the way. They die from hyperthermia, hypothermia, other illness or acts of violence. In the desert on a summer day, temperatures can soar to 120 degrees and flash rainstorms can produce instant rivers. Ironically on the same day that one migrant might die of hyperthermia–dehydration and heat-exposure, another could die in the same terrain from hypothermia–exposure to cold and wet. I arrived in July and there had been 51 recorded deaths in the month of June alone. The statistic would double or triple if it included the bodies of people who had perished in places so remote that they were never found.

The Sonoran desert is heart-wrenchingly beautiful at times. The sky is so broad and clear, the mountains peppered with flowering cacti and craggy canyons shaded by silver oaks. Living in it for one month and hiking across its wild topography gave me a vivid sense of just how treacherous this migration is. I had to drink water constantly and if I ran out towards the end of a hike I quickly lost energy and got a headache. Medical experts have evaluated that the average adult needs to drink 17 litters of water a day in these conditions to stay healthy, making it practically impossible to carry enough water to maintain proper hydration for even one day. Depending on their route and their luck, migrants are walking between 4 and 10 days. We were counseled to assume that any migrant that we met on the trails would be either moderately or severely dehydrated.

Grassroots Groups Working For Family Farmers

by TOM LASKAWY

NATIONAL FAMILY FARM COALITION
FAMILY FARM DEFENDERS
FOOD FOR MAINE'S FUTURE
VIA CAMPESINA
NATIONAL FARMERS UNION
FEDERATION OF SOUTHERN COOPERATIVES

(you can make your own decision about the farm bureau, but here is something)

New report: The Farm Bureau not a true friend to farmers

The cover of a Farm Bureau brochure.
The subtitle reads: The voice of agriculture.

The Washington Post says that the current drought is the worst in a half-century and the corn harvest will likely be even smaller than the USDA's recently downsized estimates. The wacky weather in the heart of commodity agriculture country reminds me of something Bob Stallman, president of the American Farm Bureau Federation (AFBF), said a few years ago in a Newsweek interview about farmers' attitudes toward climate change. It read: Stallman says most farmers aren't worried. "We are used to dealing with extreme weather variation," he says, pointing out that his Texas farm has seen 20 inches of rain in a single day, in the middle of a drought. "We've learned to roll with those extremes. If it gets a little more extreme down the road, we can deal with it." Yeah. Well, I don't know how many farmers want to "roll with" this drought. Mr. Stallman, would you like to revise those remarks? I'm guessing no. After all, the AFBF was instrumental in exempting agriculture from the ill-fated climate bill back in 2009 and Stallman has shown no interest in revisiting the issue. There remains no mention of climate

change on the AFBF website. But Stallman's climate change denialism is really a sideshow for the AFBF. A new investigative report by Ian Shearn for the Nation (produced in collaboration with the Food & Environment Reporting Network*), explores the full reach of the federation. It's one that encompasses political activities like lobbying and grooming state legislators sympathetic to its interests as well as vast business interests that have turned what was once a support network for family farmers into large industry players' most potent ally.

As Shearn details, it's a much more complex organization than most people realize. A key to understanding the AFBF and its state-level incarnations is the fact that the federation has 6 million members in a nation with only 2 million farmers. But Shearn explains that the federation isn't entirely what it seems: It's not just a non-profit "farmers organization" but a multi-billion dollar network of for-profit insurance companies, the third-largest insurance group in the United States. Its premiums generated more than $11 billion last year alone, on top of assets worth more than $22 billion. In many states, Missouri among them, members of the Farm Bureau board and the board of its affiliated insurance company are one and the same, sharing office buildings and support staff. And those 6 million farmers it claims as members? In many states, anyone who signs up for Farm Bureau insurance becomes a member of the Farm Bureau automatically. In Missouri, less than a third of its members are farmers. Nonetheless, all of its 113,000 members pay annual dues, as they do throughout the country, which fuels a potent political machine.

I should point out that the various farm bureau insurance companies sell all kinds of policies – not just crop insurance. But they do sell crop insurance – and lots of it. Shearn reports that, nationwide, the various farm bureau insurance companies made a combined $300 million in premium revenue in 2011. That number puts the AFBF's vigorous support of the current farm bill, which radically expands the federally subsidized crop insurance program, in a different light.

Keep in mind that the government doesn't just subsidize farmers' premiums, thus making insurance more affordable to them. The federal government also provides "reinsurance," or insurance for insurance companies, thus minimizing the companies' exposure to losses — such as you might see during a devastating drought. And thanks to the feds' financial support, there's not much incentive in there for the farm bureaus to reassess the financial risks of planting drought-sensitive commodity crops in a changing climate. It's either a virtuous or a vicious circle, depending on your perspective.

Shearn also spoke with an actual small farmer named Rolf Christen who tried to enlist the aid of the Missouri Farm Bureau in challenging a nearby 80,000-head confined hog operation that was polluting the air and water in his backyard. Rather than help the farmer, the farm bureau sided with the big operation — owned by pork industry giant Smithfield — and went so far as to get state laws changed to keep farmers like

Christen at bay.

In the end Christen went to court and won a settlement — but he claims little has changed at the offending farm. And now there's another 140,000-head hog operation just a few miles away from him. For decades, the AFBF and its state-level brethren have undertaken a dedicated effort to make super-sized livestock operations legal and welcome in states such as Missouri. And as Grist recently reported, despite all of their literal and figurative issues, the numbers of new concentrated animal feeding operations (CAFOs) in the Midwest are once more on the rise.

While you can't give the AFBF all the credit, it's been instrumental in laying the groundwork for the growth of these massive (and massively polluting) operations. It claims to be the "unified national voice of agriculture" but when pressed to choose between industry giants like Smithfield and small farmers like Christen, the AFBF backed the big guys. And it hasn't looked back.

Fig. 10.—*Water Barrel.*

ALPHABET OF RULES.

Attend well to your business.
Be punctual in your payments.
Consider well before you promise.
Dare to do right.
Envy no man.
Faithfully perform your duty.
Go not in the paths of vice.
Have respect for your character.
Interfere not with the rights of others.
Judge no man.
Know thyself.
Lie not, for any consideration.
Make few acquaintances.
Never profess what you do not practice.
Occupy your time in usefulness.
Postpone nothing that you can do now.
Quarrel not with your neighbor.
Recompense every man for his labor.
Save something against a day of trouble.
Treat everybody with kindness.
Use yourself to moderation.
Vilify no person's reputation.
Watchfully guard against idleness.
Xamine your conduct daily.
Yield to superior judgment.
Zealously pursue the right path.
&—moral—you will be happy.

NC

By The Quail

A PORTRAIT OF THE FARMER
AS AN OLD MAN

by JOHN KNIGHT

His hands are still big, even if most of the strength has been wrinkled out. These days they are rarely as dirty as they've ever been and those fingers do not always find the things they put down. But to see them clean a gun or skin a squirrel or pull down the branch of a peach tree to check the buds, you know these are hands familiar with their work.

My grandfather is ninety-six years old and his work is that of a man who lives alone on the same 1000-acre Kentucky farm where he was born in 1916. Every day he reads the Wall Street Journal front to back, drinks a tall glass of buttermilk with lunch, and goes for a walk before dinner.

Since his wife of seventy years passed away in 2005, he has been living mostly on microwaved venison and TV dinners. Sometimes he watches CNN while he eats. Sometimes he doesn't. On Sundays he dresses in a coat and tie and drives two miles to the Methodist church where he is one of five worshippers, all over the age of sixty, all born and raised in Christian County.

My aunt, his daughter, frets over him. She lives in New Hampshire with her family but would move without hesitation into the farm's cottage to watch over her father if he would let her. But he won't, even though he walks slowly and with a cane. Even though his handwriting is shaky illegible. Even though he can only hear out of one ear, and even then just barely. Fragile with age, my aunt says, thinking of ways to surround him with safety. Yet I have no doubt that anywhere else—someplace like a modern home with easy heat, no cobwebs, and a climate that doesn't freeze, or (the unimaginable) in some kind of as sisted living arrangement—my grandfather would never have celebrated his ninetieth birthday. But the farm might see him to one hundred.

A century of life I can only imagine in terms of what has changed or been lost. For my grandfather, it all might be adequately delineated by way of the Bobwhite Quail. About the size of two fists, these birds make their nests on the ground, hidden in the hedgerows that separate one field from the next. They are particular, cautious creatures that do not dally when there is not enough food or cover. When he was a boy, my grandfather spent his days with his best friend, Bill Crenshaw, gigging frogs, trapping raccoons, squirrels and foxes, and in the fall, hunting quail. The pelts they'd trade for another box of shells, and the birds they'd eat with white bread and potatoes.

These tales swim in my mind along with so many others. Like the time my grandfather's uncle ordered him to feed a sow turpentine to cure its hiccoughs. Or when he and Bill tried to impregnate their blue tick hound by throwing her into the backyards with other hounds while she was in heat. Or when they skinned a skunk, sold its pelt in town, and were suspended from their one-room schoolhouse until they could get the stink off. Or how the milk tasted when it came in from his father's cows every morning. It all sounds fantastic to my ears, a version of America that is firmly in the past whose only remnant are people like my grandfather, who eats his cereal with half and half because whole milk from the store tastes like water. In the same way that I fondly remember Pac-man from my childhood, my grandfather recalls hunting quail.

You can take the temperature of a farm by measuring any single element. The corn yield, for instance, indicates the fertility of the soil, the organization and care of the labor, the agreeability of the weather, and the difference between this year and last. A farm is a system of symbiotic pieces—the crops, the soil, the machines and the farmers all suffering, reaping, and perishing in the same drop. Productivity is not the measurement of a good farm, only an indicator of one.

My grandfather measures his farm by the quail. He says that when he was a boy, there were more than fifty coveys on this land, each with ten or twelve birds. There were also more forests, more families, and more crops. I have little reason to doubt him:

My first real memory of this man and his farm puts us somewhere around my twelfth Thanksgiving. A day before the feast, my grandfather handed me his .22 smiling. "If you can hit a quail with that thing," he said "you'll be the best shot in the family."

I remember walking out along the long strips of brambles and tall grass between the fields. I remember the dogs, two springer spaniels racing ahead. I remember the snapping carcasses of wheat and corn beneath my feet, the soil resting in frost after so many months of toil. I remember my grandfather holding his gun in the crook of his arm like he had never done anything else his whole life. I remember how still the dogs became when they smelled them and I imagined little birds crouched wide-eyed and knowing. I remember my grandfather's whistle, the rush of dogs, and the explosion from the bush. The shotguns loud but not scary, the .22 like a toy. The birds, so fast, instantly just specks in the sky, and I fired without aiming and fired again just to show that I was shooting.

Afterwards we walked to where my grandfather and uncles' spray of lead pulled three birds back down. I saw one in the grass, still alive just before a dog picked it up and delivered it to my grandfather who discreetly but without hesitation broke its neck in his big hands.

I don't remember eating the birds (though I'm sure we did). I don't even remember Thanksgiving dinner or the trip home. Just the bird in his hands and that its death was a matter of fact not taken lightly, nor considered unusual. Things live and die on a farm; the land a place of growth and harvest, its inhabitants all a part of who and what it becomes. Though I was shocked to learn this in the routine twist and snap of the bird, I also remember my grandfather's disappointment at the end of the day when we had only succeeded in flushing two more coveys, with no more than seven birds in each. In his lifetime American agriculture has been through a meat grinder, painfully and dramatically abandoning small, family-owned full diet farms to behemoth agricultural producers, managing tens of thousands of acres of a single crop. In 1910, there were 6.4 million farms in the country. Today there are about 2 million, and they average three times the size of those from a century ago.

Not only has this had a drastic affect on how the country's food is produced and by whom, but also on the price of land. With so much invested in large, expensive equipment to meet high quotas, farmers must work thousands of acres to break even. Higher demand, higher price. Combined with the increasing generational exodus from rural towns in the breadbasket to anonymous cities, the overall method of large-scale food production has left fewer and fewer families with proprietorship of any land at all. The price per acre has become too tempting for removed generations, and the soil too valuable for large-scale producers to resist buying. How valuable? Last year a 150-acre farm abutting ours sold for $10,000 an acre.

For the past thirty years, the fate of this land has been my grandfather's greatest concern. Not simply because he it's his, but because he has seen the devastation that large-scale farms have heaved upon his home. Many of the forests he roamed as a boy are now mazes of corn and on hot summer days the pesticides will sting your eyes. This Kentucky land breeds farmers notorious for cutting down and mowing anything that stands between them and a few more bushels. There are fewer and fewer hedgerows, and my grandfather hasn't seen a Bobwhite in years.

In an effort to thwart such abusive terrestrial fate, my grandfather gathered all of his land into a family corporation twenty years ago. Now we all own stocks, and when he dies, we will only be able to trade our stocks with the company or each other, rather than selling off bundles of pristine farmland. Each stock is worth $100, and there are only 5,000 of them.

A joke compared to the value of the land. There's a good chance that incorporation will prevent us from doling out and selling off little plots, though it's hard to say for sure. All around my grandfather, the families he knew as a boy have disappeared, their ghosts taunted by children and grandchildren squabbling with lawyers and each other. And truly, his own family is spread far and wide, none of us with any substantial agricultural qualifications and little real desire to live the rest of our lives in Hopkinsville, Kentucky. Oddly it's not my farm—it's my grandfather's. But even as I write this I can hear his voice in my head, reminding me: "If you've got land, boy you better not let it go, 'cause after you do, you'll never get it back."

About two thirds of my grandfather's farm is planted in high-yield row crops—corn, wheat, and soybeans. He rents it to two farmers who manage and oversee all tilling, planting, and harvesting. The other third of the land is old-growth forests, places that have never seen saw nor plow. There are oak trees governing these groves that sprouted before this country existed.

As the fall settles in, these trees give up their leaves once more, painting them with dashes of gold and crimson before letting go. My grandfather gathers the pecans that fall in his yard, and the apples from his orchard. My uncle comes and shoots a deer that they clean and package away in dinner-size portions for him to slowly eat for a year. He negotiates hunting rights with gunmen from Alabama and Arkansas. The wood is chopped and stacked, the preserves stand at the ready.

In the winter he writes letters, oils his guns, and bakes pie. His daughter and her family come for Christmas. He looks over the farm's records. Always, on his large cluttered desk in a corner of the kitchen, he keeps aerial photographs of his land. I have seen him late at night staring at them in the lamplight, trying to decipher something. He leans back with a sigh and looks up into the dark corners of the ceiling.

In the spring he plants his garden – tomatoes

and broccoli and sunflowers, the only things the deer won't eat. He watches his orchard from the kitchen window and prays that a late frost wont destroy his chances for peaches. In April he does the taxes, by hand, and hosts the annual shareholder's meeting, attended by my mother, my aunt, and myself. He organizes his contracts with farmers. He asks them about what they're planting and where. He drives his truck through his forests, looking for trees that didn't survive the winter, which he'll drag out and sell.

In the summer he watches the corn grow and picks blackberries from the groves that dot the edges of the woods. He picks his peaches and cans them, giving so many away in hope they'll come back inside a pie. He trims his hedge and mows his lawn. At the end of the day, after dinner, he sits on his front porch—the wrap-around kind so typical of southern farmhouses—and watches the robins catch worms in the yard. I am with him one time and we are silent for a while, letting the unbearable heat of the day sift off through the trees. Then he turns to me and asks what birds I hear. I tell him. Sparrows, finches, mockingbirds, robins, and a whippoorwill. "A whippoorwill, huh?" clearly pleased. "I was afraid they were all gone." A pause. "You don't hear any quail do you?"

"No," I say, though I'm tempted to lie. "But I'm sure they're there. I'll let you know if they decide to speak up."

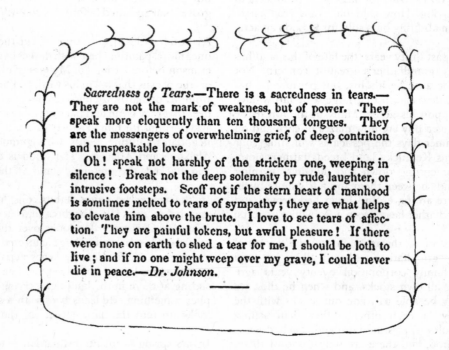

Sacredness of Tears.—There is a sacredness in tears.— They are not the mark of weakness, but of power. They speak more eloquently than ten thousand tongues. They are the messengers of overwhelming grief, of deep contrition and unspeakable love.

Oh! speak not harshly of the stricken one, weeping in silence! Break not the deep solemnity by rude laughter, or intrusive footsteps. Scoff not if the stern heart of manhood is somtimes melted to tears of sympathy; they are what helps to elevate him above the brute. I love to see tears of affection. They are painful tokens, but awful pleasure! If there were none on earth to shed a tear for me, I should be loth to live; and if no one might weep over my grave, I could never die in peace.—*Dr. Johnson.*

THE

YOUNG FARMER'S MANUAL.

VOLUME II.

CHAPTER I.—INTRODUCTION.

GENERAL MANAGEMENT OF A FARM.

"Work, for the night is coming;
　Work, through the morning hours;
　Work, when the dew is sparkling,
　Work, 'mid springing flowers."

1. A FARMER'S destiny is to labor with his hands. To facilitate his labors, and enable him to succeed in his employment, he needs *facts*. The minds of the great mass of working men have not been enlightened by scientific knowledge. For this reason, they are not properly qualified to avail themselves of the advantages to be derived from instructions that are more theoretical than practical. A well-established fact will overthrow the most plausible theory. In the management of all kinds of domestic animals, in the cultivation of the soil, in raising the various products of the farm, facts, figures, and plain details of whatever is to be done, is always of primary importance. Farmers are required to understand such a variety of manual labor, that they often need minute details to enable them to perform what sometimes appears so simple as to need no explanation. Knowing that success depends on the general management of farming operations, my aim is to aid practical men in beginning correctly and ending successfully.

14   THE YOUNG FARMER'S MANUAL.

29. When some young city dandy, who has plenty of money, starts up to be a farmer; and, with his varnished and fancifully-stenciled implements of husbandry, with his barrels of ground bone, bags of phosphate and guano, and with his retinue of Tims and Pats, just from the Emerald Isle, none of whom, boss or laborer, knows any more about either theoretical or practical agriculture than they do about the practical part of wax work; they make such droll steerage in everything they attempt to perform, that people laugh in their sleeve and say: " That's scientific agriculture !" Illustrious stupidity ! There is not one half as much correct scientific agriculture in all of that parade and counterfeit cultivation of the soil as there is of the science of medicine in a cat's eating catnip to cure the hydrophobia. It is a misnomer to call such manœuvring scientific agriculture. Again : people often see some intelligent, *theoretical* farmer commence agricultural operation; and his knowledge, for the most part, is only *theoretical.* Of course in executing the details of his practice, he as well as those in his employ will work awkwardly and unskilfully. They smile at such management and exclaim : " That's scientific agriculture !" But the wrong word is used in the wrong place. This is nothing but *theoretical* agriculture. There is none of the *scientific* about it.

WHAT IS SCIENTIFIC AGRICULTURE ?

30. Scientific agriculture consists neither in theory or practice alone; but in combining the best theory with the most approved practice in farming operations. Scientific agriculture then involves—not knowing how to analyze soils, but such a knowledge of them as will enable a man to adapt the most suitable crops to each particular soil; and also a correct understanding of the most approved manner of performing all the operations connected with raising crops, as well as of keeping the soil in a good state of fertility with the materials that the farm affords. See this thought more fully elucidated in the Chapter on Manures. This last consideration involves a correct understanding of the principles of breeding, rearing, and fattening all kinds of domestic

THE YOUNG FARMER'S MANUAL. 15

animals, of saving and applying their manure to the soil, and of cultivating the soil and raising grass or grain. All these considerations, taken in harmonious combination, constitute the sum total of scientific agriculture. It matters not if a farmer's library consists of only the Holy Bible and the Babes in the Woods, if he has a correct understanding of keeping his soil in a good state of fertility with only the materials which it affords; and if he raises good crops, and secures them well, that farmer's practices may be denominated *scientific agriculture*. Correct theory and correct practice, in the operations of a farm, make up the *scientific*. But, theory or practice alone will not do it.

31. The *Genesee Farmer* says: "It must be confessed that a man may have a good theoretical knowledge of agriculture, and yet make a poor farmer. Order, system, personal attention to details, with steady, persistent industry, will enable a farmer to succeed, without the slightest acquaintance with science; while, on the other hand, the most thorough scientific education will be of little use to the man who has not these qualities. If a man, who has had the advantage of a scientific agricultural education, turns farmer, he will be pretty sure to make mistakes which will subject him to the ridicule of his neighbors. He may be the most quiet of men—be entirely occupied with his own affairs— interfering in no way with those of others. But no matter. Those of his neighbors who have less to think about will be sure to talk over all that he does, and their comment will not generally be of a complimentary kind. Agriculture is slow work. A farm cannot be brought into order and a high state of cultivation in a year. It is the labor of a life."

32. The scientific man who thinks that he can take a farm and raise large crops by the use of a few chemical manures, is doomed to disappointment. He will be very apt to neglect those little details of farm economy which are absolutely essential to success. While he is thinking of acids, alkalies, of nitrogen and phosphates, his cattle will knock down a fence and eat up his crops. While he is studying Liebig, his men will be taking a *siesta* in the hayfield. Careless hands will soon break his im-

16   THE YOUNG FARMER'S MANUAL.

proved implements.   He may think to economize food by cooking
it; but without constant surveillance, his men will waste more
in a day than he can save in a week.   They will take pleasure in thwarting all his pet plans; and will harass and perplex
him in every conceivable way.   The end is disappointment and
disgust.

33. But it is only the mere dabbler in science that expects
to revolutionize agriculture.   The true scientific man has moderate
expectations.   He does not know, and never expects to know,
how to transmute iron into gold, or to raise a hundred bushels
of wheat per acre, as easily as we now raise ten.   If any dis-
covery he can make, if any modification of present practices will
increase the productiveness of the soil five bushels per acre, he
knows that he would be one of the greatest benefactors of his
race.   Theory can never resist facts that have once been fairly
established.   *Facts* are the leaves of science.   *Theory* alone is
not scientific.   And yet it is too commonly conceded, that if a
man theorizes much he is a very scientific man.   *Moore's Rural
New Yorker* says : "The farmer who is governed in his system
of farm management by the most extended experience and practical observation of the relation of facts to each other, is the most
scientific farmer, no matter whether he ever read a scientific book
or not.   The breeder of sheep, or cattle, who is governed in his
breeding by laws which his experience has brought out for him,
is really a *scientific breeder*, no matter whether his practice con-
flicts with the rules of theoretical writers or not."

A SUCCESSFUL FARMER MUST BE A CONSUMER.

34. Consuming the productions of the soil on the farm where
they were raised, is a practice which lies at the foundation of
scientific and successful agriculture.   Indeed, it is the only correct system of farm management which is worthy of universal
recommendation.   It is true there are numerous instances of good
agriculture, where everything that a soil produces is transported
to some other part of the country.   But even in such instances,
a system, which is *equivalent* to a consuming system, is adopted.

THE YOUNG FARMER'S MANUAL.   17

Near many of our populous cities every article that the soil has produced during the season is removed many miles distant, and consumed, where no portion of it is ever returned to the soil. But in such instances the proprietors are careful to enrich their soils with fertilizing materials such as ground bone, poudrette, guano, gypsum, lime, or stable manure.

35. But the large proportion of the farmers of our country are too remote from market to transport a very considerable portion of the products of their soil to some place where it may be profitably consumed. For this reason, if a farmer would keep his soil in a good state of fertility from year to year, so that it will be quite as productive—or more so—when he has ceased to be proprietor of it, the only correct and feasible system of management, which will enable him to accomplish such an object, will be to consume everything that his farm produces, so far as may be practicable, at home, and return the refuse and waste material to the soil. If a man is as intelligent as he should be, these suggestions will enable him to adopt some satisfactory system of management, which will be quite as profitable for him in dollars and cents, and infinitely better for his soil. Read the Introduction to Vol. I.

36. The grand and leading idea on this subject is to use up the products of the farm in such a manner as will render the same profits to the proprietor that he would receive were he to sell them when those products were in a different condition. We want to sell the products, and at the same time make such a disposition of them that the soil will not be impoverished by removing crops from it. We want to keep our cake and eat it too, in a certain sense. For example: if a farmer raises one hundred bushels of Indian corn, his aim should be to use it up in such manner that his soil will not be impoverished. The same is true of his other crops of cereal grain and grass. By feeding out one hundred bushels of Indian corn in the most economical manner, and to the best kind of swine, cattle, or sheep, and by saving all their manure and applying it to the soil where the corn grew, and by cultivating that soil in a most thorough manner, its fertility

18   THE YOUNG FARMER'S MANUAL.

may be improved, and rendered more productive from year to year.

PURCHASING A FARM—ARE YOU A PRACTICAL FARMER?

37. If you are not, whether you possess an abundance of capital to purchase a farm with, or not, my first advice is that you spend a year or two on some farm, and gain a correct knowledge of the business before you engage in it. There is something else to do on a farm than to walk around the fields with gloves on, and a staff in hand, and see the crops grow. There is a vast amount of hard labor to be done. And if a man is not able to perform much manual labor, he had better not be a farmer. If a man invests his money in mortgages or notes, he will usually receive his income without much thought or anxiety. But when it is invested in land that must be cultivated, the proprietor must be a good farmer, in order to succeed well, in obtaining an annual income which will be a fair equivalent for the use of capital invested and for the service performed. And, in addition to this, he must keep his property from depreciating in value. Otherwise, he will soon "come out of the little end of the horn," as we Americans say of a failure in any enterprise. Thousands of men have fancied that farming is a delightful occupation; and have engaged in it without knowing any more about its details, and the successes and reverses attending it, as a livelihood, than a common farmer knows about navigating the ocean. Consequently, they soon failed to meet with success, simply because they did not understand the business.

38. If you are a practical farmer, and if you like laborious and active employment, and can say *aye* to the description of a farmer in Vol. I., page 14; and if you have come, deliberately, to the conclusion to like agriculture—hard or easy, wet or dry, cold or hot, great pains with small gains, and *vice versa*—then you are fully prepared to put on your toga and India rubber boots, and work your passage to earthly bliss through swamps and bogs, hedges and ditches, as well as among stones, stumps and heaps of compost. And if you always keep cool and never fret, glorious success and substantial happiness will eventually

THE YOUNG FARMER'S MANUAL. 19

encircle you, like a halo of agricultural glory. The next important consideration, when about to purchase a farm, will be

SUFFICIENT CAPITAL.

39. There is no other occupation in which capital is important than in agriculture. If a farmer has a limited capital, he will often find farming to be disagreeable and up-hill business. He cannot expect to succeed without a fair amount of capital, any more than a banker or a tradesman who employs but little in his business. Some cash will be essential, in addition to good stock and a good soil. What I mean to have understood is, that a farmer should have capital enough on hand to defray his expenses, and a little more. There will be some improvements to be carried out almost every season. Laborers must be paid as they go ; the family must be supported ; and a man frequently meets with excellent opportunities to purchase stock for his farm at a great bargain. For all such things, a few hundred dollars of loose capital will be requisite, in order to good success. A farmer's *main* capital will be invested in his farm, implements of husbandry, and in his live stock. But if he has no loose capital besides these, he will too often find himself in a position like that of a stranger in a strange country whose property and cash are many hundreds of miles distant from him. A speculator never thinks of obtaining his livelihood by buying and selling without first obtaining a suitable amount of capital with which to transact his business. A farmer, in order to succeed, needs capital just as much as a speculator.

PAY DEBTS AS YOU GO.

40. At some seasons of the year, cash appears to be, necessarily, paid out much faster than it comes in. When this is the case, it will be infinitely better for a man——better for his credit —better for his integrity—better for his creditors—better for his family——and better in every other point of view——to hire fifty or a hundred dollars for a few months, for the purpose of paying small bills—little debts—as he goes. These little five and ten cent

20   THE YOUNG FARMER'S MANUAL.

bills at the blacksmith's shop; tnese little bills at groceries, which
cannot be avoided conveniently, had better be paid at the time
they are contracted, even though it were necessary to pay interest
for a few months on the money to do it with.   It is not a good
practice for a farmer to allow every stroke of a blacksmith's hammer to be entered on his book.   But few men will keep a book
account for nothing from year to year.   And if those with whom
we are accustomed to deal always understand that a man deals
in cash, in *little* debts as well as in large ones, dealers will allow
that man advantages in prices which a slack paymaster can never
avail himself of.

41. When I see a man go to a blacksmith's shop to get a
tenpenny job done; or to a grocery to trade a few dimes, who
always says " you may charge this bill; " and who calculates to
pay when his creditors feel willing to take their pay in catskins,
or in a quarter of poor beef, or a leg of some imbecile old buck
that is too old for mutton, I always think, My dear sir, that sys
tem of management renders your business more than ten fold
harder, and more expensive and perplexing, than it would be
were you to pay your little debts as you go.   But fools and silly
people can never be made to believe it.

PAY AS YOU GO.

42. Let me reiterate it in the ears of every young man,
whose aspirations ever prompt him to be anybody in the world,
to pay your *little* debts as you go.   Little things make the man.
Little traits of character enable us to decide with great certainty
what are a man's cherished affections; and in little things we discover manifestations of his ruling passion. The world will measure
a man by his attention to little things—by his promptness in paying his *little debts*.   The world will weigh a man's character—not
by what they *hear* about him—but by what they *see* in him. The
world will speak of a man as they find him, with reference to his
manner of doing business.   Among the undesirable and repug-
nant traits of character, this one, of allowing little debts to go
unpaid, discloses a trait more ignoble than we can discover

THE YOUNG FARMER'S MANUAL.   21

in him whose habitual weakness is to debase himself, by drinking intoxicating beverages to such an extent as to disgrace and ruin his character.

43. Again, let me repeat and reiterate for the last time, " *pay as you go.*" It will make a man a better citizen to pay all his little debts as he goes. He will esteem *himself* more highly for it; and that trait of character will secure for him a degree of respectability, in every sphere where he may move, that he never can experience if he adopts the loose, dishonest, unbusiness-like, slip-shod kind of a way of paying little debts only when they *must* be paid. Rest assured, my young reader, that your character as a successful cultivator of the soil will depend, in a far greater degree than you may imagine, on your readiness, your promptness and punctuality, in paying your *little* debts.

44. Pay as you go, young man. Pay your *little* debts. Never let it be said of you by any one : " That man owes me a dime, or a quarter of a dollar; and I would rather lose it than to ask him for it." Pay your day laborers every day, or every week at longest. They will work much better for it. Pay all little debts at mechanics' shops as soon as the work is done. They will deal cheaper on account of it. Let it be interwoven with the cardinal virtues of morality and religion, to pay every incidental indebtedness with promptness, with cheerfulness, and without murmuring and grumbling. If a man is not faithful in these *little* transactions, we can cherish no hope that he will pay his indebtedness to the soil which he cultivates, by returning to it, in fertilizing material of some kind, an equivalent for the crops which he has removed.

45. Pay as you go. For the world is full of unprincipled men, sailing under the agricultural flag of the United States of America, who run a dishonorable career, and lead a wretched life, simply because they never *calculate* to pay as they go. They are always ready and anxious to borrow a dime or a dollar, whether they have immediate use for it or not. And they never make calculations to pay it.

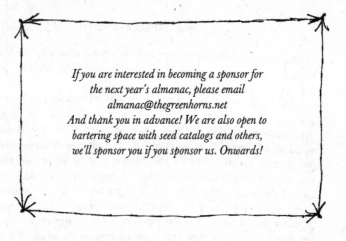

If you are interested in becoming a sponsor for the next year's almanac, please email almanac@thegreenhorns.net And thank you in advance! We are also open to bartering space with seed catalogs and others, we'll sponsor you if you sponsor us. Onwards!

SPONSORS

———

CONTEMPORARY & ARCHIVAL

OURGOODS is a barter network for the creative community.

Futurefarmers

FARM HACK

ctnofa
Local and Organic Since 1982

Living Lands
Agrarian Network

Local food.
Local farmers

Farmscape Ecology Program

SIERRA CLUB
FOUNDED 1892

Cumberland Chapter

The Beehive Design Collective
cross-pollinating the grassroots

green string farm

NATIONAL YOUNG FARMERS' COALITION

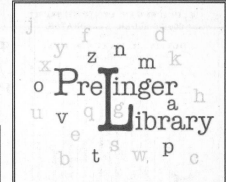

FARMERS WANTED!

WORLDWIDE OPPORTUNITIES ON ORGANIC FARMS, USA

PRODUCES A DIRECTORY OF MORE THAN 1600 ORGANIC FARMS, COMMUNITIES, & GARDEN PROJECTS IN THE UNITED STATES FOR VISITORS TO PARTICIPATE IN A CULTURAL & EDUCATIONAL EXCHANGE. WWOOFERS SPEND ABOUT HALF A DAY ON A HOST FARM, LEARN ABOUT SUSTAINABLE AGRICULTURE, & RECEIVE ROOM & BOARD -- WITH NO MONEY EXCHANGED BETWEEN HOSTS & WWOOFERS.

WWOOFERS:

- LEARN PRACTICAL FARMING SKILLS
- TRAVEL & EXPLORE THE U.S.
- EXPERIENCE LIFE ON ORGANIC FARMS
- BE A PART OF THE REAL FOOD MOVEMENT

HOSTS:

- TEACH ABOUT ORGANIC & SUSTAINABLE AGRICULTURE
- PARTICIPATE IN A CULTURAL EXCHANGE
- PASS ON VALUABLE FARMING SKILLS
- RECIEVE WILLING VISITORS TO HELP LIGHTEN THE LOAD

WHERE? ALL 50 STATES, INCLUDING ALASKA, HAWAII, THE US VIRGIN ISLANDS, & PUERTO RICO.

WHAT? A CULTURAL & EDUCATIONAL EXCHANGE FOCUSED ON ORGANIC FARMING.

WHEN? YEAR ROUND FROM A WEEKEND TO A SEASON. VARIOUS OPPORTUNITIES ARE AVAILABLE ON MORE THAN 1600 FARMS.

HOW? SIGNUP ONLINE -- PAY A SMALL ANNUAL REGISTRATION FEE, BECOME A WWOOF-USA MEMBER, RECEIVE INSTANT ACCESS TO THE HOST DIRECTORY, & CONTACT FARMS YOU ARE INTERESTED IN VISITING.

(415)-621-FARM // WWW.WWOOFUSA.ORG

NOFA-NY

NORTHEAST ORGANIC FARMING ASSOCIATION OF NEW YORK

NOFA-NY's Beginning Farmer Programs exist to offer sustainable farming education and support programs for aspiring, beginning, advanced-beginning and transitioning-to-organic farmers. NOFA-NY hosts an online directory of farm apprenticeships, connects beginning farmers with mentors and technical consultants from the sustainable farming community, plans training events with aspiring and beginning farmers' needs in mind, hosts farming-education conferences and catalogues online and in-person training programs. NOFA-NY seeks to equip any aspiring or beginning farmer with the information, resources, network of peers and support they need to feel confident and well-informed at their particular phase of starting to farm.
NOFA chapters in other states in the Northeast partner up with NOFA-NY for a concerted regional effort towards this goal.

Northeast Organic Farming
Association of New York, Inc.
(NOFA-NY)

249 Highland Ave.
Rochester, NY 14620
(585) 271-1979 ext. 511
newfarmers@nofany.org
www.nofany.org/bfam

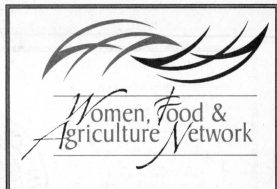

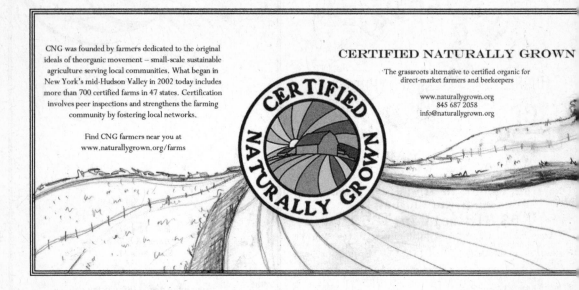

CNG was founded by farmers dedicated to the original ideals of the organic movement – small-scale sustainable agriculture serving local communities. What began in New York's mid-Hudson Valley in 2002 today includes more than 700 certified farms in 47 states. Certification involves peer inspections and strengthens the farming community by fostering local networks.

Find CNG farmers near you at
www.naturallygrown.org/farms

CERTIFIED NATURALLY GROWN

The grassroots alternative to certified organic for direct-market farmers and beekeepers

www.naturallygrown.org
845 687 2058
info@naturallygrown.org

CERTIFIED NATURALLY GROWN

18 REASONS
ART COMMUNITY FOOD

— THE —
CONVERSATIONS
WE START
AT BI·RITE,
WE CONTINUE
AT 18 REASONS

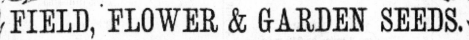

Acreage Guide

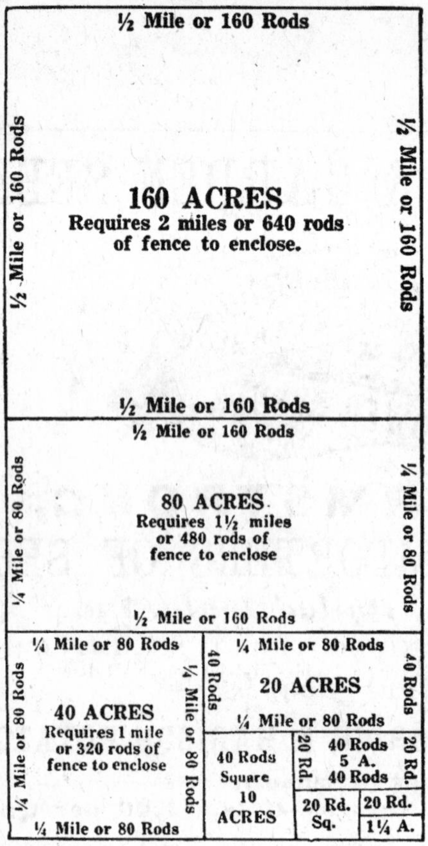

½ Mile or 160 Rods

160 ACRES
Requires 2 miles or 640 rods
of fence to enclose.

½ Mile or 160 Rods (left side)
½ Mile or 160 Rods (right side)

½ Mile or 160 Rods

½ Mile or 160 Rods

80 ACRES
Requires 1½ miles
or 480 rods of
fence to enclose

¼ Mile or 80 Rods (left)
¼ Mile or 80 Rods (right)

½ Mile or 160 Rods

¼ Mile or 80 Rods | ¼ Mile or 80 Rods

20 ACRES

¼ Mile or 80 Rods

40 ACRES
Requires 1 mile
or 320 rods of
fence to enclose

¼ Mile or 80 Rods (left)
40 Rods
¼ Mile or 80 Rods

40 Rods Square 10 ACRES	20 Rd.	40 Rods 5 A. 40 Rods	20 Rd.
	20 Rd. Sq.	20 Rd. 1¼ A.	

¼ Mile or 80 Rods

LIQUID CONVERSION CHART

1 US pint (pt)
=
2 US cups (c)
=
0.125 gallons (gal)
=
0.5 US quarts (qt)
=
16 US fluid ounce (fl. oz.)
=
96 US teaspoons (tsp)
=
32 US tablespoons (tbsp)
=
473.176473 milliliters (ml)
=
0.473176473 liters (l)

ACREAGE APPLICATIONS

Acre & Area & Acreage
by Verosocial Studio
Geo Measure - Map Area & Distance Measurement
by ObjectGraph LLC

WE

Bread & Puppet Press, Glover, VT 05839

farmersalmanac farmersatmanac farmersalmanac farmersalmanac farmersalmanac

ABOUT THE GREENHORNS

Greenhorns is a five year old grassroots organization that works to support new farmers in America. Our work is unconventional and various, we focus on event organizing, in person-networking, mixers, celebrations and workshops as well as the production of traditional and new media: radio, documentary film, blog, a book of essays, guidebooks, web-based tools. Our goal over the next generation is to retrofit the food system and to build a thriving agricultural economy, for healthy regions, healthy watersheds, and a healthy farm culture. We believe our movement can succeed with strong communication, solid business skills, sustainable farm practices and importantly, teamwork.

We hope you will join the network!
www.thegreenhorns.net to sign up.

VISIT OUR AUDIO ALMANAC

newfarmersalmanac.org